The Reciprocal Modular Brain in Economics and Politics

Shaping the Rational and Moral Basis of Organization, Exchange, and Choice

The Reciprocal Modular Brain in Economics and Politics

Shaping the Rational and Moral Basis of Organization, Exchange, and Choice

by

Gerald A. Cory, Jr.

The Center for Behavioral Ecology
San Jose, California

Springer Science+Business Media, LLC

ISBN 978-1-4613-7152-6 ISBN 978-1-4615-4747-1 (eBook)
DOI 10.1007/978-1-4615-4747-1

Originally published by Kluwer Academic/Plenum Publishers, New York in 1999
Softcover reprint of the hardcover 1st edition 1999
10 9 8 7 6 5 4 3 2 1

A C.I.P. record for this book is available from the Library of Congress

Preface

The present work is an extension of my doctoral thesis done at Stanford in the early 1970s. In one clear sense it responds to the call for consilience by Edward O. Wilson. I agree with Wilson that there is a pressing need in the sciences today for the unification of the social with the natural sciences. I consider the present work to proceed from the perspective of behavioral ecology, specifically a subfield which I choose to call interpersonal behavioral ecology

Ecology, as a general field, has emerged in the last quarter of the 20th century as a major theme of concern as we have become increasingly aware that we must preserve the planet whose limited resources we share with all other earthly creatures. Interpersonal behavioral ecology, however, focuses not on the physical environment, but upon our social environment. It concerns our interpersonal behavioral interactions at all levels, from simple dyadic one-to-one personal interactions to our larger, even global, social, economic, and political interactions.

Interpersonal behavioral ecology, as I see it, then, is concerned with our behavior toward each other, from the most obvious behaviors of war between nations, to excessive competition, exploitation, crime, abuse, and even to the ways in which we interact with each other as individuals in the family, in our social lives, in the workplace, and in the marketplace. It is about more, however, than just damage control, adjustment, and repair to the structure and behavior of our interpersonal lives. It seeks to go further – to understand and apply the dynamics of interpersonal behavior with a view to improving the larger social, economic, and political systems that shape our lives.

This present book seeks to identify and explore the basic algorithms of our evolved brain structure that underlie our social behavior and trace their dynamic, shaping effect on our social, economic, and political institutions. In the quest to unify the social with the natural sciences, we must inevitably turn to evolutionary neuroscience as the bridging discipline. There is no where else to go. Although our brain evolved under constraints of the laws of physics and chemistry, the evolutionary process itself involved chaotic and random factors, as well as natural selection processes. The algorithms of our brain, then, which are the foundation of our social sciences, can never have the immutability and predictability of the laws of physics and chemistry. These algorithms, although dynamic, shaping factors of our behavior, are, to a degree, innately variable and experientially modifiable. This fundamental difference between the natural and social sciences precludes a simplistic reduction, but indicates, rather, the establishment of linkages and bridges.

Acknowledgments

The ideas presented in this book evolved over a period of a half-century during which I accumulated many intellectual debts. My early interest in the brain goes back to my employment at Tidewater Hospital, a private psychiatric institution (1950-1951) in Beaufort County, South Carolina. Under the supervision of psychiatrist, A. K. Fidler, I participated extensively in patient care and studied the greats of psychiatry and brain science. I have pursued my interest in neuroscience unflaggingly since that time. I also wish to acknowledge Robert H. Wienefeld, former chair of history at the University of South Carolina, who, during my undergraduate years in the early 1950s helped me to begin thinking historically and economically.

I wish especially to thank Kurt Steiner of Stanford University for his guidance and encouragement. His support of the ideas and concepts of this work has been of great value to me over the years since we first met at Stanford in 1970. Also I wish to thank Robert North, Nobutaka Ike, Hans Weiler, Peter Corning, Donald Kennedy, Charles Drekmeier, Tilo Schabert, Alexander George, Richard Fagen, Daryl Bem, and Robert Packenham, all of Stanford University, who gave their advice and support in the development of these ideas many years ago. I owe special and more recent thanks to Paul D. MacLean of the National Institute of Mental Health, for his review, helpful comments, and continuing encouragement; and to Elliott White, now emeritus of Temple University, for his review of the present manuscript, his encouragement, and his many valuable comments. I also wish to thank Edward O. Wilson of Harvard University for his acknowledgement of my work and his encouragement.

For their review and endorsement of an earlier, related work, I wish to thank Alice Scheuer, of University of Hawaii (and the WHO Field

Psychiatric Center), Brian Barry, Department of Psychology, Rochester Institute of Technology, Andre Fiedeldey, Department of Psychology, University of Pretoria, Arthur Sementelli, Decision Sciences, Cleveland State University, and Trudi Miller, Department of Political Science, University of Minnesota.

Vic and Adrienne Hochee, John Lightburn, David McKenna, and Peter Lynch were my valued colleagues over many years and provided their criticisms and support with generosity and thoughtfulness. My thanks also go to my many seminar colleagues and students, who were an inspiration and a challenge over the years.

Michael Hennelly, senior editor at Plenum, was my patient guide and supporter through the production of this book.

As appropriate, I remain responsible for any errors or misinterpretations that may appear in the present work.

Permissions to quote beyond fair use have been granted as follows: Dahlman, Carl J. 1980, *The Open Field System and Beyond*, by Cambridge University Press; Gloor, Pierre. 1997. *The Temporal Lobe and the Limbic System*, by Oxford University Press; Williamson, Oliver E. 1975, 1983. *Markets and Hierarchies: Analysis and Anti-Trust Implications*, by Simon & Schuster; Hayek, Friedrich A. 1991. *Economic Freedom*, by Basil Blackwell.

Contents

Chapter 1

Introduction

This book is an effort at bridging disciplines. It responds to the recent call by sociobiologist Edward O. Wilson for consilience(1998), a concerted effort toward unifying the natural and social sciences. In the social sciences, it follows the lead of economist Gary Becker and sociologist James Coleman, who, while on the faculty at the University of Chicago, initiated a seminar aimed at the bridging of their respective disciplines (Zupan 1998).

The essential discipline to span the chasm separating the natural from the social sciences is neuroscience, specifically the study of the human brain. Although all things begin with the laws of physics, the evolutionary process that produced the brain interjects random and chaotic elements that deny prediction on a simplistic reductive basis. The human brain, then, is a product of a long period of evolution, the end product of which could not have been predicted from knowledge of the laws of physics and evolution itself. In full recognition of the complexity of the evolved brain, Ramon y Cajal, famed neuroscientist of the late 19th and early 20th centuries, reportedly said that we can never understand the universe until we understand the human brain which created that universe.

Accepting neuroscience as the bridge between the natural and social sciences does not involve a reductionist program that shrinks all social science down to fit the terms of physics and biology. Social science scholars have rightly resisted the suggestion of such simplistic reduction. The bridge of neuroscience establishes the anchors and the linkages for unification. The social sciences build upon these anchors and linkages, extending them and introducing entirely new and necessary variables from their unique perspectives and levels of analysis. Although the social variables will not be emphasized in this book, this lack of emphasis in no way denies the importance of such variables. A full discussion of the social variables is simply beyond the scope of the present work.

This book builds upon and expands three papers: the first delivered at the 1997 annual meeting of the American Political Science Association, Washington, D. C., August 28-31, 1997; a second delivered at the 1998 annual meeting of the American Sociological Association in San Francisco, August 21-25, 1998; and a third delivered at the inaugural meeting of the Society for the Multidisciplinary Study of

Consciousness, in San Francisco, August 17-18, 1998. The three papers examined the applicability for sociology, economics and political science of new concepts in cognitive science, evolutionary psychology, and neuroscience, combining them with the insights of the earlier Maslow need hierarchy and MacLean's triune brain concept.

Chapter 2 states the theme of the active, performing human organism within the context of the modular view of the computational brain tied to insights of evolutionary psychology. Within this current approach, followed by such authors as Pinker, Restak, Edelman, et al. I seek to integrate the more vintage insights of psychologist Abraham Maslow and neuroscientist Paul MacLean. The overall effort is to further clarify and define the performative, shaping dynamic that proceeds from our evolved modular brain structure to influence our social, economic, and political lives and institutions.

Chapter 3 evaluates MacLean's concept of the triune brain which has been unjustly criticized, misrepresented, and misunderstood by some critics, I assess these criticisms in detail and find them to be false, exaggerated, and often lacking in substance. I conclude that MacLean's concept is soundly grounded in evolutionary neuroscience and, with some clarifications, provides the most useful concept we have for linking neuroscience with the larger, more highly generalized concepts of the social sciences.

Chapter 4 sets out a new modular model of behavior called the conflict systems neurobehavioral model. This model, of interconnected and distributed modules, draws upon some of Maslow's insights while building primarily upon the work of Paul MacLean. Two master inclusive modules are proposed. The first consists of self-preservational programming and is based primarily, although not exclusively, in the tissues of earlier brain structures which were common to early amniotes, reptiles and mammals. The second consists of affectional programming which is based primarily, although not exclusively, in the tissues of the brain which became highly developed with mammals and produced the peculiarly mammalian affectional behaviors of maternal nursing and long-term infant-parent-family bonding. These two master modules are driven by our cellular and body processes of metabolism mediated by hormones, neurotransmitters, and neural architecture. They act dynamically to shape our behavior in the environment and set us up for a life of internal as well as external conflict because of their often conflicting priorities. Behavioral tension occurs when the two are in conflict or frustrated in behavioral expression. The often conflicting urges of these two master modules are input by way of bi-directional and multi-lateral neural pathways to the more recently evolved neocortex, which is capable of language and thought. These urgings are represented in the neocortex (a master language, thought, and executive module), and are expressed at a high level of cognitive generalization as ego (self-preservation) and empathy (affection). The executive function of the neocortex (especially the frontal cortex) has the capacity and the responsibility for making our rational choices from among these often conflicting behavioral priorities. The documentation of these three master modules draws upon a wide spectrum of neuroscience, ethology, and psychology (cognitive, motivational, and attachment).

Chapter 5 develops the algorithms (evolved rules of procedure or function) of reciprocal behavior from the tug and pull of ego and empathy. Three ranges in a spectrum of behavior are described: the egoistic, dominated by self-interested, self-preservation priorities; the empathetic, dominated by other-interested, affectional priorities; and the dynamic balance range, in which the priorities of ego and empathy are approximately balanced. These proposed algorithms, although individualized by genetic, gender, learning, and experiential differences, nevertheless constitute the neural architecture common to all human beings. Although they operate very imperfectly, these reciprocal algorithms allow us how to get to reciprocity through conflict in our behavioral motives and actions. These algorithms are seen to have a shaping effect on all interpersonal behavior, from the simplest social interactions to the most complex. The proposed algorithms are statistical, like the second law of thermodynamics and quantum physics, in that they do not allow prediction of precise behavior at the basic unit of analysis, the individual, molecular, or subatomic level respectively...but only on the basis of statistical probability. They also function analogously to a quantum wave function.

Chapter 6 contrasts the new model with the Maslow hierarchy demonstrating, contrary to Maslow, that conflict not emergence is most definitive characteristic of human individual and social behavior.

Chapter 7 traces the ubiquitous and pervasive norm of reciprocity through extensive anthropological and sociological literature. Reciprocity is the pervasive norm of social or economic exchange in any society. Balanced and unbalanced reciprocity are seen to be directly connected with inequality and equality. Reciprocity, then, which is considered by many scholars to be the sine qua non of society, is shown to express the reciprocal algorithms of behavior proposed in chapter 5.

Chapter 8 examines the role of empathy in economics. It briefly traces the evolution of the modern free enterprise market from the family or group bond, through gift-giving, and the development of the gift exchange, and the market transaction. Empathy, or other-interest, is seen to work in tandem with ego, or self-interest, to make the market possible. Empathy gives the crucial capacity to enter into and identify with the self-interest of others, making market exchange possible. The algorithms of reciprocal behavior are shown to be the dynamic of the market. Power is likewise seen as driven by the same algorithms. Differentials in capacity to provide or withhold resources equate to inequalities of power.

Chapter 9 examines the self-interested assumption of received rational choice theory and finds it to be inadequate. The reciprocal algorithms of our evolved brain structure are shown to provide a more adequate underpinning for the issues of choice and reciprocity in social exchange theory as well as its variant of equity theory. The algorithms, then, provide the linking dynamic between neuroscience and theories of social exchange, rational choice, and equity theory.

Chapters 10 and 11 examine the role of the reciprocal brain in the management and creation of scarcity from the perspective of political economy. Human society, to include the economic and political aspects, is a product of the human brain interacting with like brains under environmental constraints. There is no other possible source. There are no social, economic, or political essences or universals existing out there in a positivist, mechanical world waiting to be discovered. The human brain functions, among other things, as a normative, evaluative, and environment-shaping organ based upon its evolved mechanisms to assure survival of the individual and the species within the existing constraints. All aspects of human society, then, are normatively based. There is no such thing as a positivist, value-free human politics, economics, or any other aspect of society.

The brain evolved as a scarcity-coping organ in a primarily kinship based foraging society where sharing or reciprocity was essential to survival and which reinforced the adaptive evolution of the mammalian characteristics of self-preservation and affection. The reciprocal algorithms of behavior are a scarcity-coping mechanism. Reinforced, however, by emphasis on the self-interested logic of a limitless productivity as an end in itself, the brain becomes a scarcity-generating mechanism. The greater part of scarcity, therefore, is self-generated and reinforced by our prevailing politico-economic paradigm. A limitless productivity as an end in itself, then, unless modified and properly managed, threatens to take us to the brink of extinction by exhausting the carrying capacity of the environment. Our institutions, our derived environmental constraints, are devised to order the reciprocity dynamic. To the extent that they provide order (regulate), and they invariably do, the institutions are political. To the extent that they impact reciprocity, and they invariably do, they are economic. Institutions, then, are concerned with ordering reciprocity in some way. In that sense they are politico-economic.

Chapters 12 and 13 examine the new institutional economics from the perspectives of Oliver Williamson and Douglass North. Although both scholars attempt to deal with the problem of cooperation within a self-maximizing paradigm, they both fall short of the mark. Beneath the assumptions of both scholars lies the implicit, unarticulated dynamic of the reciprocal algorithms of our evolved brain structure. This shaping reciprocal dynamic is driven by behavioral tension. Transaction costs, added by institutions to the process of exchange, can be linked to and understood in terms of behavioral tension, thus establishing further the dynamic link between brain science and economics. The costly paradox of transaction cost economics is that its emphasis on self-interested opportunism as fundamental has the effect of increasing rather than decreasing such costs.

Chapter 14 deals with the contrived, normative definition of demand and supply in received economic theory. As defined, demand and supply represent a truncated, not a full expression of the algorithms of reciprocal behavior. They are definitions prefabricated to produce predictable theoretical results according to the received paradigm. Demand, for instance, consists of two elements, taste and the ability to buy. It therefore excludes the most relevant political and social information and conceals the most significant failures of the market. Those who need most desperately...the children, the poor, the homeless... are excluded from the definition of demand because they lack the ability to buy. A cleared or perfect market, when demand and supply curves cross at the point of price equilibrium, fails to tell us whether all, some, or even any of the children have shoes, or other necessities of life.

Chapter 15 develops the reciprocal equation that underlies behavior, social, and economic exchange. This equation, in its various forms, is the fundamental equation underlying social and economic exchange relationships, cost/benefit analysis, internal reciprocity, power relationships, and hierarchy. The encompassing equation indexes the behavioral tension of the society, the inequality, and the social and political tensions. And it demonstrates mathematically the fundamental dynamic linking of neuroscience or evolved brain structure with the social exchange sciences.

Chapter 16 reveals the culture-bound nature of received Anglo-American economics. In addition to being a nonscientific, normatively defined theory, the Anglo-American version exists alongside alternative successful versions of capitalist enterprise that emphasize different fundamental assumptions about human nature and the economic process.

Chapter 17 examines the merging of economics and political science into political economy characterized by the importation of economic rational choice theory into political science under the rubric of public choice. Despite the advantages of uniform theory and methods bridging the two disciplines, there is also a downside. The preexisting problems and distortions are carried from economics into the public choice literature based on the exclusive self-interest motive and the failure to recognize the reciprocal nature of all exchange and choice. The clarifying effect of acknowledging empathy as the reciprocal of self-interest in political theory is discussed. The addition of empathy avoids the pessimism of a Hobbesian approach and allows for a better accounting of the issues of political legitimacy, social cohesion, and social justice. It also avoids the troubling implicit academic endorsement and propagation of a one-sided self-interested egoism in public affairs. As even Kenneth Arrow, Nobel prize winner and acknowledged creator of social choice theory, has commented: "People just do not maximize on a selfish basis every minute. In fact the system would not work if they did. A consequence of that hypothesis would be the end of organized society as we know it." This book can be seen as an attempt to bring our social, economic, and political theory into line with this insight. Hopefully, with the bringing down of the overemphasis on self-interest and the acknowledgement of the balancing role of empathy, the last vestiges of Social Darwinism will begin disappear from our social, economic, and political thought.

Chapter 18 concludes that the reciprocal brain is, indeed, the dynamic, shaping mechanism across the multidisciplinary spectrum from evolutionary neuroscience through the alternative social perspectives of anthropology, sociology, economics, and political science. The combining of an ancestral (protoreptilian), self-preserving tissue complex interconnected with a mammalian nurturing, other-preserving complex, overlaid and interconnected with a massive, generalizing cortex, which adds rational, cognitive capacity combined with language, defines our essential humanity. From the dynamic of this reciprocal brain derives the reciprocal algorithms of behavior. These algorithms are the shaping dynamic of human social organization and permeate that organization from all perspectives ...anthropological, sociological, economic, and political. Their shaping influence is absolutely pervasive, although variable and probabilistic. The implications for our emerging global society are indicated.

Chapter 2

The Maslow Hierarchy of Needs vs. MacLean's Triune Brain

Political scientist, Elliott White, in acknowledging the end of the empty organism perspective that had prevailed for the greater part of the 20th century in the form of behaviorist psychology and focusing on the importance of neurobiology as a necessary foundation for the sciences of human action, to include the science of politics, writes: "A science of human life that ignores the brain is akin to a study of the solar system that leaves out the sun."(1992:1). Primatologist Shirley Strum and social scientist Bruno Latour in their article "Redefining the Social Link from Baboons to Humans,"(1991), argue for a performative model of social interaction in which society is continually constructed or performed by active social beings.

More recently, cognitive scientist Steven Pinker of Massachussetts Institute of Technology, in *How the Mind Works* (1997), brings together the computational theory of the mind and the emerging discipline of evolutionary psychology with emphasis on information processing as the primary function of an evolved, adaptive modular brain. The concept of the brain, as a set of information-processing modules evolved independently to cope with specific adaptive problems as set out by Pinker and others, has become the standard for cognitive psychology. Cognitive science, combining the insights of evolutionary psychology, seeks to discover how the mind works by, in Pinker's words, "reverse engineering." That is, it identifies adaptive behaviors in the evolutionary environment and then back engineers them to postulate specific modules in the brain to deal specifically with the identified environmental challenges. Although care must be exercised to avoid the obvious easy-way-out fallacy of postulating a separate and dedicated brain module for each behavior, function, or emotion identified, the modular view is supported by research upon the brain itself. Cognitive neuroscience tunnels into the problem from the opposite side to discover the specific brain modules or neural structures that control these functions and behavioral responses.

The cognitive science approach, coupled with cognitive neuroscience and combined with evolutionary psychology, has progressively filled the empty organism and has given us the theoretical and empirical foundations for an active, performing organism. These approaches see not the largely blank slate brain of erstwhile behaviorism, but a brain chock full of interconnected modules designed for coping with environmental challenges. Owing to the overemphasis on cognition as information processing there has been a somewhat belated growth of a complementary literature that covers or seeks to include the neglected area of feelings, emotions, and the innate reward and response systems or modules that give pure cognition or information processing its subjective quality, value, or affective meaning (e.g., LeDoux 1996, Restak 1994, Edelman 1992).

This book carries on this theme of the active, performing organism and can be seen as an effort to further clarify and define the performative, shaping dynamic that proceeds from our evolved modular brain structure to influence our social, economic, and political lives and institutions. It attempts further to integrate the findings of these new approaches with the earlier influential and more vintage insights of psychologist Abraham Maslow and neuroscientist Paul MacLean.

Models from the psychological (to include neuropsychological) sciences, however, have seldom been widely applied to issues of social, economic, and political theory. Among the reasons for this lack of applicability is that psychological models usually focus on the individual and are constructed at a level of generalization and analysis that makes them unsuitable for theorizing at the higher altitude or level of generalization of these social science disciplines. The Maslow hierarchy of needs and the triune brain concept of MacLean have both been with us for a long time (Maslow's hierarchy for the better part of five decades, MacLean's for the better part of three) and are generally familiar.

MASLOW'S HIERARCHY

In Maslow's theoretical structure, needs are usually organized from bottom to top in the form of a staircase, or stepladder as follows: physiological needs (hunger, thirst), safety needs, belonging or social needs; esteem needs; and the self-actualizing need. Maslow theorized that these needs were emergent: That is, as we satisfied our basic needs of hunger and thirst, our safety needs would then emerge. As we satisfied our newly emerged safety needs, the next level, the belonging or social needs, would come into play. Next came esteem needs, and finally, as these were satisfied, the self-actualizing need at the top of the hierarchy emerged (Maslow 1943, 1970, 1968).

Maslow's hierarchy has appeared in every basic text on psychology and behavior for the past four decades. It also appears in most texts on organizational behavior. Its influence has been widespread as a behavioral scheme of ready and easy reference. It has also been popularized in casual and impressionistic writing about motivation. Maslow's well-known concept represents one of the earliest comprehensive efforts to develop a model of the human biological inheritance.

The Maslow hierarchy has, however, serious shortcomings that limit its utility for conceptualizing the genetic inheritance. For one thing, it lacks an evolutionary perspective. The hierarchy of needs is presented as a given, disconnected from the evolutionary process which produced it. Secondly, the concept of hierarchy is not fully developed. It does not allow sufficiently for interaction of the levels of hierarchy and does not account for those cases that violate the normal priority of needs (Cory 1974: 27-29, 85-86; Corning 1983: 167-72; Maddi 1989: 110-118;

Smith 1991). Maslow's hierarchy has also been criticized for being culture bound, fitting neatly with particularly the U.S. concept of material achievement and success as a steady stairstep progression of higher development (Yankelovich 1981). It thereby tends to ignore or diminish the great accomplishments in thought, morality, and service to humanity of many of the great figures of human history (Maddi 1989). Maslow's hierarchy, with its almost exclusive focus on the individual, affords little insight into the dynamics of social interaction. In its rather long history, despite some attempts, it has failed to become a major influence in socialization or political theory (Zigler and Child 1973: 33-5; Knutson 1972:168-72, 261-3. Davies 1963, 1991, has made the most consistent effort to apply Maslow's concepts to politics).

MACLEAN'S MODULAR CONCEPT: MISMEASURED AND MISUNDERSTOOD

MacLean's triune brain concept is one of the earliest modular concepts of the brain. Although it has been acknowledged to be the single most influential idea in brain science since WorldWar II (e.g., Durant in Harrington 1992: 268), it has largely been overlooked by cognitive psychology. In an extreme case it has been summarily and undeservedly rejected as wrong by Steven Pinker in his recent book noted earlier. This anomalous situation, in which the pioneering modular statement of brain organization coming from neuroscience itself and providing a natural match with aspects of the modular cognitive approach, has been brought about by a couple of seriously flawed reviews of MacLean's work that appeared in the influential journals *Science* (1990) and *American Scientist* (1992).

The effect of these faulty reviews has been to deny the use of MacLean's very significant research and insights to the researchers in the cognitive psychological as well as the social science community, who relied upon the authority of these prestigious journals. In fact Pinker bases his unfortunate and mistaken rejection of MacLean's thought solely on a reference to the review in *Science* which is the most prejudicial and grossly inaccurate of the two (Pinker 1997: 370, 580). The detailed and documented rebuttal of these reviews is reported in chapter 3. The presentation that follows here is adjusted to accommodate criticisms where valid.

MacLean sees behavior as essentially irrational, motivated and validated by earlier nonverbal brain structures. In fact, this irrationality is a frequent theme of concern to MacLean (1990, 1992). Accordingly, in the few instances when MacLean's triune brain concept has been applied to society and politics, this factor of irrationality has been given major play (e.g., Peterson 1981, 1983; Pettman 1975: 153-75). This emphasis on irrationality, however, has obscured the potential value of MacLean's concept. Extended, elaborated, and applied thoughtfully, it provides the neuroscientific basis for a better understanding of the structure and dynamics of our social and political lives.

THE INTERCONNECTED, THREE-LEVEL (TRIUNE) BRAIN

In a recent thoroughgoing, encyclopedic summary of the last fifty years of brain research, MacLean (1990) documents the human brain as an evolved three-level interconnected structure. This structure comprises a self-preservational, maintenance component inherited from the stem reptiles of the Permian and Triassic periods, called the protoreptilian complex, a later modified and evolved mammalian affectional complex, and a most recently modified and elaborated higher cortex.

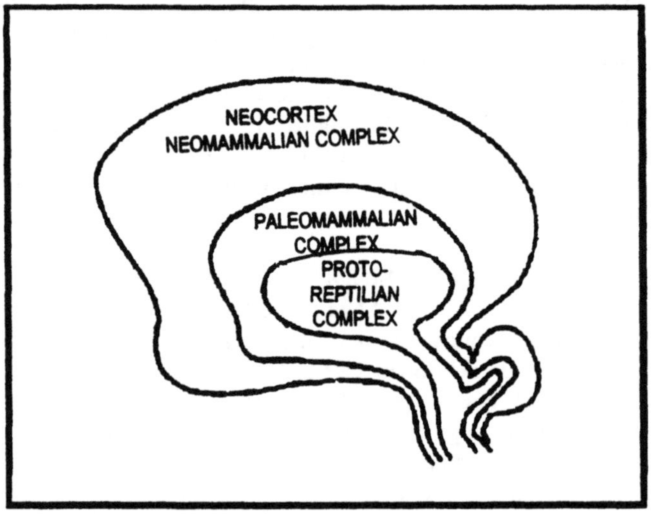

Figure 1. A simplified, modified sketch of the triune brain structure (After MacLean). As represented here the three brain divisions do not constitute distinct additions but rather modifications and elaborations of probable preexisting homologues reflecting phylogenetic continuity.

As brain evolution progressed in the line ancestral to humans, simple protoreptilian brain structure was not replaced, but provided the substructure and homologues for subsequent brain development while largely retaining its basic character and function. Accordingly, the brain structure of early vertebrate life forms ancestral to humans (i.e., early fishes and reptiles) became the substructure and provided the homologues for the mammalian modifications and neocortical elaborations that followed and which have reached the greatest development in the brain of humankind. Appreciating the qualitative differences of the three levels is important to understanding the dynamics of human behavior.

The protoreptilian brain tissues in humans are proposed, as they did in the ancestral stem reptiles, to govern the fundamentals, or the daily master routines, of our life-support operations: blood circulation, heartbeat, respiration, basic food-getting, reproduction, and defensive behaviors. Such functions and behaviors were the essential routines also to be found in the ancient stem reptiles. Located by MacLean in what are usually called the hindbrain and the midbrain (i.e., the brain stem) as well as in certain structures at the base of the forebrain, this primal and innermost core of the human brain makes up almost the entire brain in fishes and amphibians.

The next developmental stage of our brain, which comes from rudimentary mammalian life and which MacLean called the paleo- or "old" mammalian brain, is identified with the structures designated collectively as our limbic system. Developing from homologues preexisting in the protoreptilian brain, these newly elaborated limbic tissue-clusters included such physiological structures as the amygdala, the thalamus, the hypothalamus, the hippocampus, and other structures. Behavioral contributions to life from these modified and elaborated paleo-mammalian structures, or limbic system, included, among other things, the mammalian features (absent in the stem vertebrates) of warm-bloodedness, nursing, infant care, and extended social bonding. These new characteristics were then neurally integrated with the life-support functional and behavioral circuitry of the protoreptilian tissue complex to create the more complex life form of mammals.

The neocortex, which MacLean called the neo- or "new" mammalian brain, is the most recent stage of brain modification and development. This great mass of hemispherical brain matter that dominates the skull case of higher primates and man, by elaborating the preexisting homologues present in the brains of early vertebrates, overgrew and encased the earlier ("paleo-") mammalian and protoreptilian neural tissues, but essentially did not replace them. As a consequence of this neocortical evolution and growth, those older brain parts evolved greater complexity and connectivity in support of these new tissue structures and in response to the behavioral adaptations necessary to life's increasingly sophisticated circumstances. The part of our brain that MacLean termed "protoreptilian" is thus actually considerably more complex than that of ancestral reptiles, the transitional therapsids of the Permian and Triassic periods, and ancestral mammals. And the part of our brain that MacLean called "paleomammalian" is also much more complex than that of ancestral and lower mammals. Each part, however, may be considered to serve, in the main, its original functions: our protoreptilian brain structures, though modified from those of ancestral reptiles, nevertheless principally regulate our basic survival mechanisms, while our paleomammalian structures, though modified from those of other mammals, principally govern our nurturing behaviors.

Since, as previously noted, MacLean's very useful three-level modular brain concept has been subjected to invalid and inhibiting criticism in some quarters of neuroscience, the next chapter is directed to addressing and countering the specifics of these criticisms so that the relevance of the concept for the social sciences can be fully established and appreciated. Those readers having no interest in the detailed criticism and rebuttal may wish to skip this discussion and go on to the next chapter.

Chapter 3

MacLean's Triune Brain Concept: In Praise and Appraisal

Paul D. MacLean is a pioneer, a trailblazer, a scientist, and thinker well ahead of his time. As a humanist deeply interested in the larger questions of human life, he started out studying philosophy. Unable to find satisfactory answers to questions such as the origin and meaning of life...why humans in spite of their unrivaled intelligence, often behaved in seemingly irrational ways threatening their individual as well as species survival...he turned to medicine and the study of the human brain. He anticipated that the brain, as the biological substrate of these behaviors, held the key to better understanding of these fundamental questions as well as hopefully their answers. MacLean was, for many years, chief of the Laboratory of Brain Evolution and Behavior of the National Institute of Mental Health. In 1952, drawing upon the nineteenth century French scientist, Paul Broca's designation of the great limbic node which surrounded the brainstem of mammals, he introduced the conceptual term *limbic system* into the neuroscientific literature. In 1970 he introduced the concept of the triune brain, which became widely popularized after the publication of Carl Sagan's rather overly dramatic and simplified discussion of it in *The Dragons of Eden* (1977). MacLean, in further developing the triune brain concept, which aroused great interest in psychiatry, education, and the lay public, produced his detailed and highly documented volume, *The Triune Brain in Evolution: Role in Paleocerebral Functions* in 1990.

CRITICISMS OF MACLEAN'S MODEL

MacLean's triune brain concept has been acknowledged the single most influential idea in neuroscience since World War II (e.g., Durant in Harrington 1992: 268). Nevertheless, following the publication of his 1990 opus, MacLean received highly critical reviews in two prominent science periodicals, *Science* (October 12, 1990: 303-05) and *American Scientist* (September- October 1992: 497-98). Both reviews were written by neurobiologists and both reviewers claimed that MacLean's triune brain concept has had limited acceptance or been largely ignored by professional neurobiologists.[1]

Anton Reiner, of the Department of Anatomy and Neurobiology, University of Tennessee, at that time a recent graduate, wrote the review in *Science*, which was the more extensive of the two. After initially recognizing MacLean as a trailblazer of neuroscience, whose triune brain concept has been well-received outside the field of brain research, as the centerpiece of Sagan's popular, *The Dragons of Eden*, and frequently as the only discussion of brain evolution in psychiatry and psychology textbooks, Reiner makes several points in criticism of the triune brain concept.[2]

First, he notes that since MacLean introduced the concept, there has been tremendous growth in neuroscientific research that has greatly extended our knowledge of brain function and evolution. This statement, of course, carries the general implication, which Reiner later makes explicit, that the concept is out of date.

Secondly, in initiating a criticism of MacLean's concept of the limbic system, Reiner writes: "MacLean's presentation of the role of the hippocampus in limbic functions is not well reconciled with the current evidence that the hippocampus plays a role in memory."(1990: 304).

Thirdly, Reiner contends that current research indicates that MacLean's reptilian complex is not a reptilian invention but seems to be present in vertebrates all the way back to jawless fishes.

Fourthly, Reiner maintains that MacLean overreaches the evidence when he claims that the basal ganglia are the neural seat for the control of species -typical types of behaviors.

Fifthly, Reiner states that the limbic system, which widely used term MacLean authored as a pioneer neuroresearcher, is not properly represented by MacLean. Contrary to MacLean, as Reiner would have it, the limbic system did not appear first in early mammals. Amphibians, reptiles, and birds also have limbic features such as the septum, amygdala, a different-looking hippocampal complex, and maybe even a cingulate cortex.

Sixthly, Reiner asserts that MacLean assigns the functions of parental behavior, which Reiner claims that MacLean regards as uniquely mammalian, to the mammalian cingulate cortex, ignoring the fact that some reptiles (crocodiles), all birds, and possibly even some extinct reptiles (dinosaurs) also engaged in parental behavior.

[1]For a highly favorable review of MacLean's 1990 book see the review by Emre Kokmen, M.D. of the Mayo Clinic, Rochester, Minnesota, in *J. Neurosurg*. V. 75, Dec, 1991, p. 998. In this chapter I focus on the reviews in *Science* and *American Scientist* because they have reached a wider audience and have become red flag reviews unjustifiably inhibiting the thoughtful application of the triune brain concept in related fields as well as in the psychological and social sciences. This chapter is a slightly modified version of an article appearing in *Across Species Comparisons and Psychopathology (ASCAP) Newsletter*, July 1998(Cory 1998).

[2]The criticisms made by Reiner are not necessarily in the exact order presented.

Seventhly, Reiner makes a couple of other criticisms of MacLean concerning a) his preference for correspondence over the more evolutionarily appropriate concept of homology and b)his apparently uncritical acceptance of Haeckel's idea that ontogeny recapitulates phylogeny.

Finally, although Reiner praises MacLean's motives and acknowledges the appeal of the triune brain concept for dealing with "big" behaviors that we are all interested in such as: "How does our animal heritage affect our behavior? Why do we do the things we do? Why can we not live together more harmoniously?"... he feels that there are some telling shortcomings as recited above, in MacLean's scholarship. He concludes that "neuroscience research ***can*** (emphasis mine) shed light" on these important human questions, "though ***perhaps*** (emphasis mine) not in as global and simple a way as MacLean has sought."

A CRITIQUE OF REINER'S CRITIQUE

Book reviews because of their very nature are usually overly brief. They usually cannot deal in depth with the points they take issue with. Reviewers, then, are often themselves guilty of the same kinds of oversimplifications and misinterpretations that they seek to expose in their reviews. When Reiner states..."I strongly believe the triune-brain idea to be wrong"... he is caught up in the same oversimplifying tendency that he claims unjustifiably to find troublesome in MacLean.

The triune-brain concept may be wrong in some of its particulars, right in others, but still be very useful and valid in its more general features. After all, at this stage of our knowledge of the brain although it is quite advanced over the 1960s and 1970s, there are not a great number of things we can say with absolute confidence...very few generalizations that are without arguable interpretations of more detailed research data. And Reiner takes apart but does not offer a replacement generalization. His analysis is destructive, not constructive. This type of analysis is the easy part of the job...almost anybody can do it.

But in his apparent eagerness to discredit and take apart MacLean's useful generalization, he also fails to study his subject closely and therefore engages in some very careless scholarship. He makes significant omissions, outright errors, and substantial misrepresentations of MacLean's work. Let's look at the points Reiner raises one by one.

1. ***Reiner blatantly misstates the facts when he claims that the triune brain concept as well as MacLean's book is outdated and lacks up-to date documentation.***

Reiner's first point i.e., that there has been a great growth in knowledge about the brain since MacLean first announced his triune brain concept in the 1960s and 1970s implies that MacLean has left the concept untouched and undocumented since that time and has therefore not considered any of the more recent findings. The implications of this statement are belied by the currency of research cited by MacLean and included in his discussions. In backing up his case for the alleged outdated ideas and data in the book, Reiner baldly states "only a handful of papers from the '80s are cited"(Reiner 1990: 305). This statement is categorically false and easily contradicted by a count of bibliographic items. The bibliography of this work contains over 180 entries (a big handful indeed!) which date from 1980 to at least 1988 and over 220 entries that date between 1975 and 1979. This amounts to at least 400 entries of rather recent documentation...keeping in mind, of course, that the publication date of MacLean's book and Reiner's review was for both 1990.

2. ***Reiner ignores or misstates the facts when he says, "MacLean's presentation of the role of the hippocampus in limbic functions is not well reconciled with the current evidence that the hippocampus plays a role in memory."***

The phrasing of this statement indicates that MacLean is unaware of or fails to report on the extensive research indicating the role of the hippocampus in memory. Such an implication is totally unwarranted. MacLean devotes fully two chapters to reporting and discussing such research. These chapters even have memory in their titles. Chapter 26 is titled *Microelectric Study of Limbic Inputs Relevant to Ontology and **Memory*** (emphasis mine). Chapter 27 is titled *Question of Limbic Mechanisms Linking a Sense of Individuality to **Memory*** (emphasis mine) *of Ongoing Experience.* These chapters deal at length with the role of the hippocampus in memory and propose an integrative role for the hippocampus in tying learning to affect or emotion (For a summary of MacLean's discussion on these matters, consult 1990: 514-16).

3. ***Claiming that the reptilian complex is not a reptilian invention, Reiner misrepresents MacLean's position.***

On the third point, Reiner contends that current research indicates that MacLean's reptilian complex is not a reptilian invention but seems to be present in vertebrates all the way back to jawless fishes. This is largely a taxonomic question. At what point do we declare something to be a fish, an amphibian, an amniote, a reptile, or a mammal? And do we view mammals as branching off from the amniote tree before we have distinct reptiles in the line of descent? Or do we prefer the more likely probability that mammals descended in a line from the ancient mammal-like reptiles of the pre-dinosaur Permian-Triassic periods called therapsids, who represent a branching of the ancient reptile line (cotylosaurs). Therapsids appeared approximately 230 millions years ago, and approximately 50 million years before the emergence of the great dinosaurs of the Jurassic and Cretaceous periods.

MacLean knows these facts and clearly acknowledges them, while supporting a lineage for mammals that traces back to the therapsids, of the synapsida subclass that branched off from the diapsida line that eventually produced the great dinosaurs many years later. This is the standard position in evolutionary theory today. One might wish to compare the phylogenetic tree in MacLean (1990: 34) with Butler and Hodos (1996: 72), Strickberger (1996: 396) and Hickman, et al. (1984: fig. 27-1). And it is the accepted position of standard zoology texts (e.g., Miller and Harley 1992, Hickman, et al. 1984, 1990). Mammals, and ultimately us humans, then, did not evolve from dinosaurs but from a parallel lineage that split much further back in geologic time.

If the term Reptilian Brain or Reptilian Complex causes confusion with modern reptiles, and because the reviewers don't wish to read MacLean's work closely, the Reptilian Complex could be thought of, and perhaps redesignated, as the ancient amniote complex or even the early vertebrate complex. And, of course, as MacLean acknowledges thoroughly, this early brain complex is not the reptilian brain of modern reptiles but it is also not the same as that of the early vertebrates, amniotes, or therapsids. At several points in his book, MacLean makes this unequivocally clear by his reference to stem reptiles (cotylosaurs) (MacLean 1990: 33, 82), those early reptiles from which both the diapsid and synapsid lines branched off. To assure the proper evolutionary context, MacLean also uses the term *protoreptilian* in his initial definition and adds the clarifying comment that he refers to the reptilian complex (or R-complex) only for brevity's sake (see MacLean 1990: 15-16, 244, 519). This protoreptilian, or stem reptile brain, has been altered by modifications which include those produced by differentiation and elaboration of earlier structures (e.g., see MacLean 1990, 243). These modifications, to include differentiations and elaborations, provide, in addition to their previous maintenance and behavioral functions, neural circuitry in support of the enhanced limbic structures of mammals. These enhanced mammalian limbic structures necessarily engage and enhance prior circuitry in the brainstem. And together these enhanced limbic and brainstem circuits provide support for the greatly enhanced neocortex (or isocortex) which eventually got the sufficient modifications that permitted language and the development of complex technological societies.[3]

4. ***Reiner misrepresents MacLean's position on the basal ganglia.***

On the fourth point above, Reiner states that he knows of no one other than MacLean who believes the basal ganglia to be the neural seat for the control of species-typical types of behaviors (Reiner 1990: 305). This statement is a misrepresentation of MacLean's position as well as an admission of ignorance on the part of Reiner. In the first place, MacLean never uses the inclusive term "neural seat." Further MacLean is not talking about all species typical behavior but only some. He specifically excludes from this discussion such mammalian class/species typical behavior as maternal nursing and play, which are attributed primarily to other brain parts and treated in other chapters of the book.

In part II on the *Striatal Complex with Respect to Species-Typical Behavior*, MacLean repeatedly emphasizes that the traditional view that the striatal complex is primarily involved in motor functions represents an oversimplification. He writes that the purpose of the present investigation is to test the hypothesis that the striatal complex plays an "essential" role in certain species typical behaviors as well as certain basic forms of behavior common to both reptiles and mammals (MacLean 1990: 243). At one point after reciting the evidence, MacLean says that the results "suggest that the medial globus pallidus (a structure of the basal ganglia) is a site of convergence of neural systems involved in the species-typical mirror display of gothic-type squirrel monkeys." (MacLean 1990: 189). And, a little further on, that "findings indicate that in animals as diverse as lizards and monkeys, the R-complex is ***basically involved*** (emphasis mine) in the organized expression of species-typical, prosematic communication of a ritualistic nature." (1990: 189).

[3]The use of the term "additions" is deliberately avoided here because it has been the source of some confusion (see Butler and Hodos 1996: 86). New brain structures do not spring de novo out of nowhere but rather evolve from the differentiation of previously existing structures. When differentiations become sufficiently established, they are often referred to loosely as "additions." This does not deny that seemingly new additions may possibly and occasionally arise, but the intent here is to emphasize the phylogenetic continuity that underpins the concept of homology

Additional research, some predating some postdating Reiner's review and of which Reiner is apparently ignorant, adds further support to MacLean's hypothesis. For example, J. Wayne Aldridge and colleagues from the University of Michigan in a research report titled "Neuronal Coding of Serial Order: Syntax of Grooming in the Neostriatum,"(1993) conclude that there is "direct evidence that the neostriatum ***coordinates the control*** (emphasis mine) of rule-governed behavioral sequences." This study builds upon a series of earlier studies of species-typical grooming behavior of the rat (e.g., Berridge and Fentress 1988; Berridge and Whishaw 1992; Cromwell and Berridge 1990). These earlier and more recent studies certainly support MacLean's hypothesis that the striatal complex plays an essential role in some species typical behaviors of a ritualistic nature.

And, of course, there is the growing body of clinical evidence, going well back into the 1970s and 1980s, that neurological disorders in humans (such as Parkinson's, Huntington's, and Tourette syndromes), which involve damage to the neostriatum, produce specific deficits in the sequential order of movement, language, and cognitive function.(e.g., Holthoff-Detto, et al. 1997; Cummings 1993; Benecke, et al. 1987; Marsden 1982, 1984; Oberg and Divac 1979). Such serial order patterns in behavior are phylogenetically old as well as pervasive and often constitute the basis of identifying so-called species-typical behaviors.

5. ***Reiner misrepresents the facts when he claims that MacLean says the limbic system first appeared in mammals.***

On the fifth point above, Reiner again misrepresents MacLean's position. MacLean does not claim that the limbic system first appeared in early mammals. He acknowledges that limbic features appear in fishes, reptiles, and birds, but are rudimentary and poorly developed as compared with those of mammals (MacLean 1990: 247, 287). According to MacLean's view, then, it is not the presence or absence of limbic features themselves in ancestral amniote or reptilian vertebrates, but rather the significant and prominent development of limbic features in mammals which is appropriately of interest in understanding the evolution of characteristically and uniquely mammalian behavior. Further, care must be exercised in making comparisons across existing modern species. We can only infer that the structures and undeveloped and/or rudimentary homologues of such structures in modern species were also present in ancestral lines. Brains don't fossilize, so the point can't be made conclusively. The currently accepted inferential position in neuroscience is that there are homologues of limbic structures going well back into vertebrate history, although these homologues in modern species are often difficult to establish and sometimes downright dubious (Striedter 1997; Veenman, et al. 1997).[4]

[4]The accuracy and utility of the concept and term limbic system has itself been a separate topic of some disagreement in recent years. Some authors state that it does not represent a truly functional system and the term should be discarded. Others defend its use. Most texts continue to find the term useful and because of its longtime usage it will probably remain in the literature. Some recent and prominent scholars illustrate the controversy well. Pierre Gloor of the Montreal Neurological Institute, McGill University, in his thorough-going work *The Temporal Lobe and Limbic System*, by the very use of the term in the title indicates his position. Further on in the text, while acknowledging the controversy he writes:

"...In addition, in mammals the hippocampus and amygdala exhibit close mutual interconnections that do not appear to be at all prominent in nonmammalian vertebrates. Thus

6. ***Reiner displays careless scholarship and misrepresents the facts of neuroscience, evolution, and animal behavior as well as MacLean's position on parental behavior and the cingulate cortex.***

Another example of careless scholarship on Reiner's part is in the sixth point above. He claims that MacLean assigns the functions of parental behavior, to the cingulate cortex and that MacLean regards parental behavior as uniquely mammalian. According to Reiner, MacLean's alleged position "ignores the fact that some reptiles, such as crocodiles, and all birds engage in parental behavior, not to mention the possibility suggested by paleontological data that some extinct reptiles, namely dinosaurs, also engaged in parental behavior."(Reiner 1990: 305).

Such a blanket claim suffices to make one wonder if Reiner felt it worth his while to even consult the book he is reporting on. Firstly, MacLean does not "assign" parental behavior to the cingulate cortex. He reports the recent (at that time) research on maternal mechanisms in the septal or medial preoptic area (MacLean 1990: 351-53) and indicates that this area may have provided the initial potentiality for full scale mammalian maternal behavior (MacLean 1990: 354), which would

---

in all mammals the hippocampus and amygdala, together with and partially through the basal forebrain areas and the preoptico-hypothalamic continuum extending along the medial forebrain bundle down to the midbrain tegmentum, form the core of a system, the limbic system, which has some anatomical and functional unity inasmuch as it embodies mechanisms that relate 'external' reality perceived by the exteroceptive senses to 'internal' reality embedded in memory and affect. This system in mammals exhibits an organization that is sufficiently different from that characterizing other areas of the cerebral hemisphere to merit such a designation."(Gloor 1997: 106).

And well-known neurologist, Richard Restak tells us that based upon a large body of experimental work, it is appropriate to conclude that, "depending on the areas stimulated, the limbic system serves as a generator of agreeable-pleasurable or disagreeable-aversive affects."(1994: 143). Nevertheless, there is little agreement among neuroscientists concerning the contributions of the different components, and their mutual influence on each other (1994: 149).

On the other hand William Blessing, a neuroscientist at Flinders University, in his study of the lower brainstem, feels that emphasis on the limbic system has detracted from the study of brainstem mechanism, that it has been "plagued by its anatomical and physiological vagueness and by the lack of precision with which the term is used."(Blessing 1997: 15). Further, he feels the term should be dropped from the literature(Blessing 1997: 16).

A third recent author, neuroscientist Joseph LeDoux (1996: ch. 4) argues that because the limbic system is not solely dedicated to the single global function of emotion, a claim that MacLean fully recognizes in his chapters on memory (1990: chs: 26 & 27), that the concept should be abandoned. LeDoux apparently prefers a single functional criterion for the definition of a system, whereas MacLean seems to prefer a combination of functional and anatomical criteria. Le Doux concludes his argument by stating: "As a result, there may not be one emotional system in the brain but many."(1996: 103). Compare this with the concluding line of the definitional description by Kandel, et al., authors of the most widely used textbook on neuroscience and behavior: "The limbic system contains neurons that form complex circuits that play an important role in learning, memory, and emotion."(1995: 708).

The use and value of the conceptual term limbic system, then, seems to depend on one's research focus and how one chooses to define a system. It might be added that the definition of what constitutes a system is controversial in all disciplines, not just in neuroscience.

include play and the development of empathy. The very title of his chapter 21 is ***Participation*** (emphasis mine) *of Thalamocingulate Division in Family-Related Behavior.* Participation is participation not unilateral and unequivocal assignment. And MacLean uses the systemic term thalamocingulate to indicate intra-limbic nuclei and cortical connections, not simply cingulate cortex as Reiner states. MacLean cites good evidence for thalamocingulate participation in "nursing, conjoined with maternal care"(MacLean 1990: 380). After all, lesions in certain portions of the cingulate cortex interfere with nursing and other maternal behavior (Stamm 1955, Slotnick 1967), not with blanket parental care as Reiner asserts.

It may be too early or simply erroneous in neuroscience to assign anything specifically and finally to any exclusive part of the limbic area. There is more likely some localization of minor function, but for most behaviors of any scale there seems to be fairly wide-ranging neural circuitry that may be interrupted by lesions at many different points. For example, recent research on maternal behavior (nursing, retrieval, nest-building) in rats has focused on the medial preoptic area with its connections to other limbic structures and the brain stem (Numan 1990). Alison Fleming and her colleagues (1996), summarize what we know about the neural control of maternal behavior. Not only the medial preoptic area with its brainstem projections, but also other limbic sites are involved, including the amygdala (Numan, et al. 1993; Fleming, et al. 1980), hippocampus (Terlecki and Sainsbury, 1978; Kimble, et al. 1967), septum (Fleischer and Slotnik 1978), and cingulate cortex (Slotnik 1967, Stamm 1955). Most emotions, emotional behaviors, and emotional memories seem to be distributed, involving multiple pathways. Specific behaviors and categories of behaviors can be interrupted by lesions at varying points in these multiple pathways. More recent research has again confirmed that the cingulate cortex is involved in emotion and motivation (Stern and Passingham 1996). In a recent research report John Freeman and colleagues conclude that the neural circuitry formed by interconnected cingulate cortical, limbic thalamic and hippocampal neurons has fundamentally similar functions in the affective behaviors of approach and avoidance (Freeman, et al. 1996).

Like any good scientist with an open mind, MacLean, at the close of his chapter on participation of the thalamocingulate division in family-related behavior, calls for more neurobehavioral research to explore the extent of this participation (MacLean 1990: 410). It is also noteworthy that MacLean is one of the few thinkers in neuroscience who shows concern for the neural substrate of such family based behavior, characteristic of mammals, as play and the underpinning but illusive quality of empathy. Although such characteristics have been reported on behaviorally (e.g., for play, see Burghardt 1988, 1984; Fagen 1981), they have largely been ignored in the search for neural substrates, not because they are unimportant, but because of the extreme difficulty in defining and objectifying them. But the evidence clearly points to neocortical as well as limbic cortical and subcortical representation (e.g., see Fuster 1997: esp. 169; Frith 1997: 98; Frith 1989: 154-55). One of these days, hopefully, mainstream neuroscience will direct more serious research toward a better understanding of these difficult and ignored questions which are so critical to a full understanding and appreciation of humanity.

Reiner also indiscriminately uses the blanket term "parental behavior" coupled with attributing that same blanket usage to MacLean. In this usage, Reiner shows a remarkable deficit of scholarship, naivete, or both. MacLean is not discussing all parental behavior. He is discussing those nurturing behaviors that are the most distinguishing characteristic of mammals and a fundamental part of their taxonomic classification and differentiation from birds and reptiles. These behaviors must be found in either new structures or modifications to existing structures. As Butler and Hodos point out, new structures may be added to organ systems, but modification of existing structures appears to be more common (1996: 86). The jury is still out on the neurophysiology of these defining mammalian behavioral features. What's more, with the emphasis on cognition in neuroscience, there has been surprisingly little attention paid to the extensive work on the neural and hormonal basis of the motivational and emotional aspects of maternal care. This is openly acknowledged by leading scholars in the brain science field (e.g., Rosenblatt and Snowden 1996; LeDoux 1997: 68; Kandel, Schwartz, and Jessell 1995).

The blanket term "parental care" as used by Reiner in his criticism of MacLean amounts to condemnation by indiscriminate generalization. Parental care has been defined by a leading authority as "any kind of parental behavior that appears likely to increase the fitness of the parent's offspring" (Clutton-Brock 1991: 8). It is a very broad and inclusive term. The term includes nest and burrow preparation. The very production of eggs itself is included. This kind of "parental care" is found in the earliest vertebrates with very primitive brains indeed. If the all-inclusive definition of parental care can be stretched to include the production of eggs and digging a hole to place them in, perhaps it could conceivably be stretched to include even the sharing of cellular membranes during asexual reproduction by single-celled organisms.

But specifically...what about parental care in modern reptiles? Contrary to Reiner's claim, MacLean reports on parental care in crocodiles (MacLean 1990: 136-37) and also in some species of skink lizards (MacLean 1990: 136, 248-249). A recent review article on parental care among reptiles by Carl Gans of the Department of Biology, University of Michigan, brings us up to date. Gans claims that the most spectacular example of reptilian parental care takes place among crocodiles. Both parents respond to the call of hatchlings who vocalize underground while emerging from the eggs. The adults dig them up and transport them to water in their large buccal pouch(Pooley 1977). The young are then washed and stay shortly in association with the adults. After a relatively brief period, however, the juveniles' response to the adults reverses. The juveniles disperse suddenly into small, nearby channels where they may dig themselves tunnels. Gans notes:

> In view of the fact that crocodylians may be ***cannibalistic*** (emphasis mine), there seems to be both an inhibition of cannibalism in the parents and an inhibition of a possible adult avoidance reaction in the neonates (1996: 153).

This kind of short-lived parental care during which the cannibalism of parents is inhibited may be impressive in reptiles, but it is a far, far cry from the highly developed family-related behavior in mammals; behavior which is so further developed in the human species that it extends often throughout an entire lifetime and becomes the basis for a vastly extended social life. The ***equating*** of parental care in reptiles with parental care in mammals is simply ludicrous. It is this mammalian family behavior that concerns MacLean, and the neural substrate is appropriately sought in the brain modifications that became prominent with the appearance of mammals.

7. ***Reiner's further inaccuracies: recapitulation, homology, and correspondence, etc.*** Near the end of his review Reiner makes the following isolated statement: "MacLean also errs in his apparent sweeping acceptance of Haeckel's idea that ontogeny recapitulates phylogeny." Again Reiner distorts and misrepresents. From a close review of the book it is by no means altogether clear that MacLean "sweepingly" accepts Haeckel's concept. In fact he only refers to it once (MacLean 1990: 46) while at the same time noting the well known exceptions. Haeckel's concept has been replaced in neuroscience today by the principles of von Baerian recapitulation. The von Baerian version holds that while ontogeny does not recapitulate phylogeny in the thoroughgoing Haeckelian sense, it does recapitulate the features of an organism in terms of the organism's general to more specific classification. In other words the von Baerian principles state that the more general features of an organism develop before the more specific features (Butler and Hodos 1996: 51-2). The issue, however, is still not so clearly settled. The emergent discipline of evolutionary developmental biology is looking more closely into such questions (Hall 1992, Thompson 1988). For instance, evolutionary biologist Wallace Arthur, in summarizing the main themes of this emerging discipline, writes:

> No single comparative embryological pattern is universally found or can be described as a 'law'. Von Baerian divergence, its antithesis (convergence) and a broadly Haeckelian (quasi-recapitulatory) pattern can all be found, depending on the comparison made (1997: 292).

On the additional point that MacLean prefers to think in terms of correspondence rather than homology probably reflects his functional-behavioral orientation. In fact it is specifically in discussing the issue of the relationship between structure and behavior that (MacLean 1990: 37) makes this comment. Later, he returns to a more standard use of homology (MacLean 1990: 228). There is, in fact, presently no sure fire way of demonstrating that homologues have the same one-to-one functions or produce the same one-to-one behaviors across species. In reporting that MacLean, at one point, expresses preference for the term correspondence because of the confusion in the definition of homology, Reiner shows what can only be considered a misplaced and sophomoric "gotcha" exuberance. He writes that MacLean's comment "should leave Stephen J. Gould, not to mention all other students of evolution, aghast," adding that such a comment constitutes a "very critical misjudgment to make in a work on evolution."(Reiner 1990: 305).

This is truly a naive, if not preposterous statement by Reiner. Could it be that Reiner is not aware of the long history of the pervasive problems associated with the definition of homology? For example, Leigh Van Valen, of the biology department of the University of Chicago, in the first sentence of his frequently referenced article on homology and its causes, writes: "Homology is the central concept of anatomy, yet it is an elusive concept."(1982: 305). Further on, in view of the persistent definitional ambiguities, Van Valen practically equates the two terms homologue and correspondence when he writes: "In fact, homology can be defined, in a quite general way, as ***correspondence*** (emphasis mine) caused by a continuity of information" ...although in a footnote Van Valen admits that correspondence itself needs further definition beyond the scope of his paper (305: fn. 1; cf. Roth 1994). Although there has been some sharpening of the concept of homology, with emphasis on phyletic continuity, the ambiguities have by no means been adequately resolved (Arthur 1997: 171-77; Hall 1994, 1996).

And there is the haunting question that is still wide open for research and investigation ...do most homologous behaviors share a homologous structural basis or can homologous behaviors be rooted in nonhomologous structures? (see Hall 1996: 29 fn. 23). The recent report by William Blessing on the lower brainstem raises the question of multiple neural representations of body parts and behavior, in that behavior originally represented and controlled in the brainstem of an earlier vertebrate may maintain its brainstem representation, but be controlled by an added representation in the frontal cortex of a more highly developed mammal. Such multiple representations at different levels as the brain became more complex would certainly confuse the issue of a straightforward homologous match of structure and function (1997: 1-18; see also, Brown 1977).

Research on very limited aspects of function are often suggestive but far from conclusive even on such limited function. Establishing homologues of the prefrontal cortex can be particularly vexing. A recent research article by Gagliardo and colleagues, "Behavioural effects of ablations of the ***presumed*** (emphasis mine) 'prefrontal cortex' or the corticoid in pigeons" (Gagliardo, et. al., 1996), indicates, not only in its discussion and conclusions, but in the very title itself, the uncertainty, ambiguity, and cautions that currently characterize such research efforts (see also Fuster 1997: 7-11).

There is an awful lot of assuming going on in some quarters of neuroscience on this issue, which simply cannot be settled at this time based on the empirical evidence. This is one of the problems and cautions that must be acknowledged when generalizing across species...say from rats to humans. In maternal behavior, for example, can we say factually that the medial preoptic area plays the same part in the maternal behavior of humans that it does it the rat brain? No, we cannot. At least not yet. But neuroscientists, after first hedging themselves, and following homologous logic, seem inclined to think so. Nevertheless, it is entirely within the realm of possibility that we may find that it does so only in part or not at all. As neuroresearcher Joseph LeDoux notes: "Some ***innate*** (emphasis mine) behavioral patterns are known to involve hierarchically organized response components." (1996: 120). And further on he adds: "Species differences can involve any brain region or pathway, due to particular brain specializations required for certain species-specific adaptations or to random changes."(1996: 123). And neurologist Richard Restak points out that in the case of animals multiple limbic areas may increase, modify or inhibit aggression. He notes further that even the same area may increase or inhibit responses under different experimental conditions and depending on the animal selected for experiment. As an example, he points out that the destruction of the cingulate gyrus (a limbic component) increases aggressive behavior in cats and dogs, whereas, on the contrary, such an operation has a calming effect in monkeys and humans (1994: 149).

Or perhaps, as Blessing suggests, there are multiple representations. Then we might have to go to correspondence rather than homology (even homoplasy might not apply, since homoplasy, or parallel evolution, would probably not apply in such closely related species) to account for the behavioral circuitry. In other words the corresponding neural circuitry--that circuitry controlling maternal behavior --may be found in the same, slightly differing, multiple, or perhaps (though highly unlikely) even totally different structural homologues or modifications.

In fact, if homology is correct and functionally, to include behaviorally, uniform...that is, the same structures account for the same functions and behaviors across classes, orders, and species... this finding would support the triune brain concept as set out by MacLean, which says generally that the protoreptilian complex common to both reptiles and mammals functions largely the same in both classes. This finding would also support MacLean's position that the expanded circuitry areas of the mammalian complex bear characteristically mammalian functions and are the circuitry for characteristically mammalian behaviors such as nursing, a defining taxonomic feature of mammals (which, in part distinguishes them from reptiles and birds).

In a final series of somewhat negatively gratuitous comments Reiner writes about some of MacLean's legitimate speculations. For example, Reiner states "...and mathematical skill (he thinks the cerebellum could be involved)"(Reiner 1990: 305).

And why not? See MacLean's discussion on the subject (MacLean 1990: 548-52). Recent research has indicated that the cerebellum is not just a motor mechanism, but is also likely involved in higher cognitive and perhaps even language function. Especially relevant is the rather well-supported hypothesis that indicates a cerebellar mechanism involved in all tasks that require precise temporal computations. This could well suggest an involvement in mathematical processes. True, the evidence is insufficient to permit firm conclusions as to the cerebellar role in higher cognitive processes, but it is a research direction which needs further refinement and is currently pursued by a number of neurobiologists (Daum and Ackermann 1995; Dimitrov, et al. 1996; Altman and Bayer 1997: esp. 749-51).

Overall, given the outright errors, careless scholarship, misrepresentations, and sophomoric, prejudicial tone of Reiner's review, it probably should never have been allowed to appear in a publication of the stature and influence of *Science.* Such prejudicial reviewing should perhaps raise serious questions of standards if not ethics in the academic-scientific community.

CAMPBELL'S REVIEW IN AMERICAN SCIENTIST

The review by Campbell in *American Scientist* (1992) is a much shorter review than that of Reiner. It brings up some of the same points, but is less prejudicial in its tone. Since it is less detailed it expresses primarily the preferences and value judgements of the reviewer. Campbell repeats Reiner's erroneous charge about outdatedness. He writes: "...that except for a very few papers, most of the references were published prior to 1980"(1992: 498). It has already been noted that this "handful" of items amounts to more than 180 citations. One suspects that Campbell proceeded from his preconceptions and found what he expected to find. Campbell ends his review with the statement: "Unfortunately, the data presented are, ***to some degree*** (emphasis mine), outdated, and the evolutionary reasoning is unsophisticated."(1992: 498). The use of the term "unsophisticated" by the reviewer is a good example of gratuitous abuse of review. It is a sweeping value-laden term that communicates more about the reviewer than the reviewed. For anyone who has closely read MacLean's detailed and thoughtful work, such blanket judgments are not warranted. The evolutionary reasoning is, on the contrary, quite thoughtful, well-presented, and sophisticated. Such blanket judgments tell us more about the sociology of neuroscience and neuroscientists that they do about the subject matter of the discipline itself.

THE COMMENTS OF BUTLER AND HODOS

In their recent comprehensive and overall admirable work on comparative vertebrate anatomy, Butler and Hodos attempt to formalize the assignment of MacLean's work to the relics of history. Their comments reflect the standard oversimplified criticisms, misrepresentations, and errors that have become popular to repeat ever more unreflectively. Butler and Hodos assign the triune brain concept, inaccurately and indiscriminately, to a category they called "theories of addition." And without any detailed discussion or analysis, of the very significant indisputable points of accuracy in MacLean's concept, they write that the past three decades of work in comparative neurobiology "unequivocally" contradicts MacLean's theory (1996: 86).

It seems almost incredible that two such qualified authors should accept the same flagrant misrepresentations, inaccuracies, and oversimplifications of MacLean's work that have become commonplace in some sectors of neurobiology over the past decade. It appears that they merely parroted the errors and misrepresentations of Reiner and others rather than reading MacLean's 1990 work closely and open-mindedly. Or perhaps they simply took their understanding from Carl Sagan's overpopularized and oversimplified presentation in *The Dragons of Eden* and didn't consider the issue worth looking into further. There is no point in repeating the responses given earlier to Reiner's review. The same points hold for Butler and Hodos' comments. The rebuttal points are clearly made and easily accessible to verification by anyone who chooses to make the effort. The categorical statement by Butler and Hodos that the extensive body of work in comparative neurobiology over the past three decades unequivocably contradicts MacLean's theory, which they apparently have not read, constitutes on that point poor, if not irresponsible, scholarship.

THE UTILITY AND VALIDITY OF MACLEAN'S TRIUNE BRAIN CONCEPT

The triune brain concept may have its faults. But such faults have been patently misrepresented in some cases and grossly exaggerated in others. Whatever its faults may prove to be, the triune brain concept gets at a fundamental truth. The mammalian modifications, differentiations, and elaborations to the early vertebrate and ancestral amniote brains had the effect of introducing endothermy (warm-bloodedness), maternal nursing, enhanced mechanisms of skin contact and comfort, as well as enhanced visual, vocal, and other cues to bond parents to offspring and serve as the underpinning for the extended and complex family life of humankind. The mammalian modifications, therefore, added greatly enhanced affectional, other-interested behavior to the primarily (although not exclusively) self-preservational, self-interested behaviors of ancestral amniotes and early vertebrates (not necessarily their modern representatives).

The simplistic representation and attempted demolition of MacLean's triune brain concept is not good science. Reiner's review, where it has any validity at all, is like discovering a termite or two in the bathroom wall -- and then proceeding to pronounce a full alarm that the house is full of termites -- only to find that it is necessary to treat a couple of boards in the subflooring. Further, in his deconstructive, analytic fervor, Reiner has not offered an alternative higher level generalization. The review represents a dysfunction common to a lot of scientific practice ... that of an analytical approach that takes apart but can't put back together. Perhaps we should call it analytic myopia. Being not interested in the bigger questions of humanity that we so desperately need help on, and lacking an interest in therapy, these analytic myopics continue their fine-grained focus. Fine-grained focus is fine, laudable, and very much needed. It becomes analytically myopic, however, when it fails to place in context what it finds and defines, when it employs sloppy scholarship, and when it attempts prejudicially to destroy or deconstruct that which it lacks the imagination and courage to put together.

On the other hand the theories of brain evolution that Butler and Hodos review favorably and the synthesis which they present at the end of their book focus on the immunohistological, hormonal, and morphological mechanics (1996: 463-73). They say, in fact, almost nothing at all about behavior or the significance for behavioral evolution for the various mechanisms of evolution they identify. And they make no attempt whatsoever to confront the larger behavioral questions of humanity where we need help and guidance from neuroscience in defining the neurobiological basis of human nature in order to establish links up the scale of generalization with the social sciences. The theories they present are only of interest to the technical aspects of neuroscience. They are not, however, incompatible, but rather tend to support MacLean's concepts when these concepts are thoughtfully considered and not inaccurately reported, misrepresented, or grossly oversimplified.

The key point in comparing these theories with that of MacLean's is that they are comparable, at best, only in part. They ask and respond to different questions. MacLean tries to address the larger questions of human nature and behavior. The others show no interest in such questions but address the fine grained technical questions of anatomical and functional evolution. At the level where they meet they do not contradict each other but are largely compatible. At the point they diverge they primarily address different questions. This is, I think, the root of the tension between the two. MacLean's concept facing up the scale of generalization is useful and has been appropriately well-received in the therapeutic sciences, and is also very useful for the social sciences. On the other hand, it has not been, but may yet become, more useful and better received in other quarters of neuroscience ...especially when subjective experience is eventually given its due in the study of consciousness. There are, in fact, recent signs that the importance of subjective experience, which is of great interest to MacLean, is beginning to be more fully recognized in the newer studies of consciousness (Hameroff, et. al. 1996).[5]

The triune brain concept may need modification, then, as the body of neuroscience grows...but certainly not outright rejection. With appropriate

[5]See especially the article by Stubenberg (1996); also, Galin (1996: 121) who writes: "I assert that what is most interesting about mental life for most ordinary people is not mechanism, not performance, not information processing; it is what it feels like! Subjective experience!" Searle (1997) provides a general criticism of the emerging consciousness literature. See also the assessment by molecular neuroscientist Smith (1996: 471-74).

clarifications, it is still by far the best concept we have for linking neuroscience with the larger, more highly generalized concepts of the social sciences. This is true even if its level of generality has limited utility for some neuroscience researchers who are doing ever more fine-grained research into neural architecture and function.

The transition from early vertebrate to amniote to synapsid reptile to mammal was in behavioral effect the transition from a nearly exclusively self- preserving organism with relatively little or less complex social life to, at least in part, a nurturing, "other-maintaining", "other-supporting", or "other-interested" organism. And that makes all the difference in the world for human evolution. Our other-maintaining mechanisms combined with our self-preserving ones provide the biological glue as well as the dynamic for our remarkable behavioral evolution, our social life, and ultimately the crucial social and political factor of our moral consciousness.

The qualitative differences between the familial and social behaviors of even the most caring of reptiles (say, modern crocodiles), birds or social insects and the mammal we call human are overwhelmingly evident. Humans with their social, cognitive, and language skills, for better or for worse, dominate the planet and no other species comes close. Any neurobiologist who cannot see or appreciate the difference is suffering from analytic myopia or some form of misplaced species egalitarianism (e.g., see Butler and Hodos 1996: 3-4). The proper study of humans is humans and to some extent their lineal antecedents. The triune brain concept generalizes a fundamental truth out of much that is yet unknown and uncertain in neuroscience. And this generalization, when properly understood, appreciated, and applied, is the most useful bridging link, thus far articulated, between neuroscience and the larger and pressingly critical questions of humanity's survival...as well as the hoped for transformation of humanity into a truly life-supporting, planet-preserving and enhancing custodial species.

When other neuroscience researchers reach the conceptual point in their grasp of the discipline where they feel an increasing obligation to take a more holistic view and proceed to move up the scale of generalization in order to confront the larger questions of human life, they will likely produce concepts closely resembling the triune brain. Homology and behavioral evolution will almost inevitably take them in that direction. When that time comes, if the triune brain concept has been buried in the scrap heap of scientific history, it will be exhumed, refurbished, and honored. Frankly, despite its current lack of popularity in some quarters of neurobiology, I do not think it will be consigned to the scrap heap. I think that it will continue to be influential, and with appropriate modifications as research progresses, provide an important underpinning for interdisciplinary communication and bridging.

Chapter 4

Toward a New Neurobehavioral Model

The unique features of the human brain evolved over a period of several million years in a primarily kinship based foraging society where sharing or reciprocity was essential to survival and which reinforced the adaptive evolution of the mammalian characteristics of self-preservation and affection.[6] Ego and empathy, self-interest and other- interest, are key features of our personal and social behavior. To relate these to MacLean's concept we need a subjective/ behavioral rather than a neuro-physiological vocabulary...one that will express what the presence of our proto-reptilian and paleomammalian brain structures mean with regard to our day--to-day, subjectively experienced, behavioral initiatives and responses to one another and the world we live in. In computer-related vocabulary, familiar to us all through cognitive psychology and artificial intelligence, I use the software designer's vocabulary of programs and programming. I will speak of our three developmental brain levels as behavioral programs or sets of programs that subjectively drive and generate specific, and objectively observable, behaviors.[7]

From the predominantly survival-centered promptings of the ancestral protoreptilian tissues as elaborated in the human brain arise the motivational source for egoistic, surviving, self- interested subjective experience and behaviors. Here are the cold-blooded, seemingly passionless, single-minded behaviors that we've

[6]For example see Isaac (1978) and Tooby and DeVore (1987). Cosmides and Tooby surmise that cognitive development in humans allowed a widening and diversification of items of social exchange (1989: 59).

[7]For earlier versions of the behavioral model developed here see Cory (1974, 1992, 1996). Also compare the model of human communication by Dingwall (1980) based in reflexive (striatal or reptilian) affective (limbic or paleomammalian), and cognitive (neocortical or neomammalian). Dingwall draws upon Lamendella (1977). See also Leven 1994.

generally associated with the present-day lizard, the snake, and that most maligned of fishes, the shark.[8]

Here is a world revolving almost exclusively around matters of self-preservation. The protoreptilian brain structures, then, will be referred to as our self-preservation programming.

From the infant nursing, care-giving, and social bonding initiatives and responses of the mammalian modifications and elaborations arise the motivational source for nurturing, empathetic, other-interested experiences and behaviors.

Here are the warm-blooded, passionate, body- contacting, bonding behaviors that we've come to identify with the lion, the wolf, the primates.[9]

Here is a world in which nearly single-minded self-preservation is simultaneously complemented and counterpoised by the conflicting demands of affection.

The early mammalian modifications, then will be referred to as our affectional programming.[10]

[8]Experimental work in animals as diverse as lizards and monkeys shows the reptilian complex is involved in displays of agonistic and defensive social communication. Also it is noteworthy that partial destruction of the reptilian complex eliminates the aggressive, territorial display (MacLean 1993: 108).

[9]The division of function between the protoreptilian complex and the limbic system is not clear-cut, but rather entangled. The lower structures of the limbic node have been shown to augment the self-preservational behavior of feeding, fighting, and self-protection (MacLean 1990; 1993: 109), adding passion or emotion to them(Kandel, et al. 1995: 595-612). The newer structures in the upper half of the limbic node, especially the septal, including the medial preoptic area, and thalamocingulate division, are involved in the affectional, family-related behavior (Fleming, et al. 1996; MacLean 1993: 109).

[10]Positing the affectional programming draws not only upon current neuroscience but also the extensive literature on the concepts of social bonding and attachment, especially the work done on higher primates and man. For fundamental work on lower animals see the pioneering work of the Austrian ethologist and Nobel prize winner Konrad Lorenz (1970, 1971). Particularly relevant here would be the work of psychologist, Harry F. Harlow on the nature of love and attachment in rhesus/macaque monkeys (1965, 1986). Harlow described five affectional systems in monkeys -- maternal, mother-infant, age-mate, heterosexual, and paternal (1986). In this paper I have proposed one all- inclusive affectional program. There has been a recent resurgence of interest in the evolutionary biological basis of affection and empathy, especially in primates (e.g., Goodall 1986; de Waal 1996). In the case of humans, the work of Spitz (1965) and British psychiatrist John Bowlby(1969, 1988) is of special interest. All the foregoing reflect field observations, experimental behavioral observations and clinical work. None of them penetrate the brain itself. More recent work in computer modeling of neural processes has focused primarily on cognition and avoided dealing with the more complex issues of affiliation and emotion. For example, Churchland and Sejnowski in their extensive and well known work on the computational brain acknowledge the neglect of these critical areas (1992: 413). From the standpoint of neuroscience, it is also notable that Kandel, Schwartz, and Jessell, authors of the most widely used text on introductory neuroscience also show this neglect (1995). Extensive research has been done on the role of the amygdala in emotion, but such research has generally focused on the emotion of fear (LeDoux 1997). The neglect is not difficult to explain. Research on such complex pathways within the brain, in spite of great progress in recent years, is still in its very early stages. The

Before I go on to discuss the neo- mammalian neocortical structures in behavioral terms, I wish to pause to consider how these first two sets of programs function together.

Our Evolved Brain and the Sources of Subjective/Behavioral Conflict.

These core behavioral program modules, composed of (or served by) sets or subsystems of modules of our brain structure, serve as dynamic factors of our behavior. They are energy-driven by our cellular as well as overall bodily processes of metabolism as mediated by hormonal, neurotransmitter, and neural architecture. Each is an inextricable part of our makeup, because each is "wired into" our brain structure by the process of evolution. Behavioral conflict exists, then, simply by virtue of the presence of these two large-scale energy-driven modular program sets in our lives -- up and running even prior to birth. Their mere physiological presence sets us up for a life of inner and outer struggle, as we are driven by and respond to their contending demands.[11] Conflict is more than an externalized, objective ethical, moral, or decision-making dilemma, however. Subjectively, feelings of satisfaction occur when we can express our felt motives, while feelings of frustration occur when either our self-preservational or affectional impulses cannot be expressed in the behavioral initiatives and responses we wish to make.

Behavioral tension then arises. Experienced as subjectively defined variants such as frustration, anxiety, or anger, behavioral tension occurs whenever one of our two fundamental behavioral programs -- self-preservation or affection -- is activated but meets with some resistance or difficulty that prevents its satisfactory expression. This subjective tension becomes most paralyzing when both programs are activated and seek contending or incompatible responses within a single situation. Caught between "I want to" and "I can't" -- e.g., "I want to help him/her, but I can't surrender my needs" -- we agonize. Whether this tension arises through the thwarted expression of a single impulse or the simultaneous but mutually exclusive urgings of two contending impulses, whenever it remains unresolved or unmanaged, it leads to the worsening condition of behavioral stress.

The Blessing of Tension and Stress.

The evolutionary process by which the two opposite promptings of self-preservation and affection were combined in us enhanced our ability to survive by binding us in social interaction and providing us with the widest range of behavioral responses to our environment.[12] Our inherently conflicting programs are a curse,

---

unknowns are still very vast. Currently the best summaries of research in neuroscience on the nurturing, caring, family-related behavior are contained in Fleming, et al. (1996); MacLean (1990: 380-410, 520-62). However, Kalin (1997), in an article focusing on the neurobiology of fear, also reports research relevant to the brain mechanisms mediating affiliation.

[11]In cognitive neuroscience brain modules are commonly seen as competing and also cooperating (e.g., see Crick 1994; Baars 1997) The idea of competing or conflicting modules contriving behavioral tension is also acknowledged by Pinker (1997: 58, 65).

[12]The evolution of the neocortex, our big brain, was in all probability greatly enhanced by the tug and pull of our conflicting programs. Humphrey (1976) sees the function of the intellect providing the ability to cope with problems of interpersonal relationships. See also the discussion in Masters (1989: 16-26). Cummins (1998) argues that interpersonal relationships,

then, only to the degree that we fail to recognize them as a blessing. Our self-preservation and affection programs allow us a highly-advanced sensitivity to our environment, as well as the ability to perceive and appreciate the survival requirements of others. Ironically, the accompanying behavioral tension -- even the stress! -- is an integral part of this useful function, for it allows us to more immediately evaluate (a subjective function) our behavior and the effect it is having on ourselves and others.

Behavioral tension serves as an internal emotional compass that we can use to guide ourselves through the often complicated and treacherous pathways of interpersonal relations.

Behavioral stress tells us that we are exceeding safe limits for ourselves and others and for our larger social, economic, and political structures.

Behavioral tension and stress are, at this point perhaps needless to say, inherently and necessarily subjective.

But of course all of this requires a certain level of consciousness, perhaps best designated self-aware consciousness, coupled with the ability to generalize our internally experienced motives. If all we possessed were the conflicting programs of self-preservation and affection, we would, of course, be among the life forms whose behaviors are governed by instinct. We would be driven by the urgings of fight, or flight, or bondedness; and every so often -- like the legendary mule who, thirsty and hungry, looked back and forth between water and hay, unable to move -- we would be caught in the indecision of those urgings.

But whether or not other mammals with paleomammalian brain structures, with self-preservation and affectional programming, experience conscious conflict from these two behavioral priorities, we certainly do. We can reflect and generalize, not only upon our choices, but also upon the meanings they have for our personal as well as our species' existence and significance. And it's in that capacity to reflect, to self-consciously experience, generalize, and decide upon the tug-and-pull of our conflicting urgings, that we come to third stage of brain development in MacLean's model: the neomammalian or "new" mammalian brain structures... what I have designated the executive programming.

THE CONFLICT SYSTEMS NEUROBEHAVIORAL MODEL (CSNM)

It's our expanded and elaborated neocortex (or isocortex) that provides us with self-aware consciousness, and with the evolutionarily unique and powerful ability to use verbal language to create concepts and ideas by which to interpret our consciousness, to describe the feelings, motives, and behaviors that arise within us and in response to our social and environmental experiences.[13]

---

competing and cooperating with conspecifics for limited resources, is the chief problem confronting social mammals. Cummins concentrates on dominance hierarchies which she sees as dynamic rather than static.

[13]A language module did not, of course, pop out of nowhere and appear in the neocortex. The capacity for spoken language involved modifications of supporting anatomical structures including the laryngeal tract, tongue, velum (which can seal the nose from the mouth) and the neural connections that tied in with motor areas necessary for the production of speech. These all evolved relatively concommitantly from the hominid ancestral line and, combined

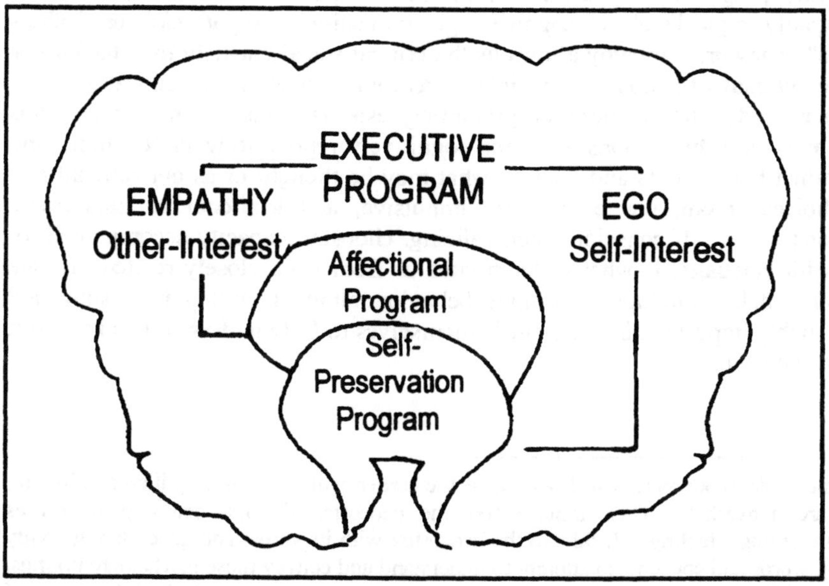

Figure 2. The conflict systems neurobehavioral model. A simplified cutaway representation of the brain showing the behavioral programs and the derivation of Ego/self-interested and Empathy/other-interested motives and behaviors. The positioning of ego and empathy is intended to indicate probable frontal lobe laterality (see Tucker, Luu, and Pribram 1995).

It is with this executive programming we acquire the ability to name, to comment upon, to generalize[14] and to choose between our contending sets of behavioral

---

with the claboration of the neocortical structures of thought and syntax, made language possible. This example of the complexity of language development provides a caveat to avoid overly simplistic one for one specialized module for specific behavioral or functional adaptation positions. The work of Philip Lieberman, a linguistic psychologist at Brown University is especially relevant for the understanding of this very complex language capability. See the up-to-date treatment of these issues in Lieberman' s *Eve Spoke* (1998).

[14]The ability to self-consciously generalize is apparently a unique gift of the neocortex with its billions of neurons interconnected into hierarchical networks. The level of generalization issue in all our disciplines likely springs from this. That is, we can move from parts to wholes in generalizing and from wholes to parts in analyzing freely up and down throughout our neural networks. Generalizing (and implicitly analyzing) has been recognized by scholars in many disciplines as perhaps the defining characteristic of the human brain (e.g., Hofstader 1995: 75; Einstein 1954: 293). This generalizing capacity loosens up the tight wiring of routines and characteristics of earlier brain structures and allows us to manage and, to some degree, overcome the mechanisms that we inherited in common with kindred species. In other words, the generalizing, analyzing capacities of the neocortex change the rules of the game for us humans by freeing us up from the blind tyranny of primitive mechanisms. This capacity must always be weighed when trying to apply findings in, for example, even primate ethology to humans. One of the reasons our feelings and motives are so difficult to verbalize

impulses: self-preservation, commonly called, at a high level of cognitive generalization, "egoistic" or "self-interested" behavior, and affection, which we call, at an equally high level of cognitive generalization, "empathetic" or "other-interested" behavior. Empathy allows us the critical social capacity to enter into or respond emotionally to another's self-interest as well as other emotional states.[15]

In other words, our executive programming, especially our frontal cortex,[16] has the capability and the responsibility for cognitively representing these limbic and protoreptilian brain inputs and making what may be thought of as our rational and moral choices among our conflicting, impulsive, and irrational or nonrational motivations. This self conscious, generalizing, choosing capacity accompanied, of course, with language, is what differentiates us from even closely related primate species and makes findings in primate behavior, although highly interesting and unquestionably important, insufficient in themselves to fully understand and account for human behavior.

---

and communicate to others is probably because the earlier evolved brain (reptilian and limbic) systems are nonverbal. Their input enters the neocortex through neural pathways as inarticulate urgings, feelings. It falls to the neocortex with its verbal and generalizing ability to develop words and concepts to attempt to understand and convey these inarticulate urgings. MacLean (1992:58) states that the triune brain structure provides us with the inheritance of three mentalities, two of which lack the capacity for verbal communication.

[15]My use of the term empathy here includes the affectional feelings of sympathy which are dependent upon empathy, plus cognitive aspects (Hoffman 1981). Losco has noted that empathy, amplified by cognitive processes, could serve as an evolved mediator of pro-social behavior (1986: 125). Empathy and sympathy are frequently used inclusively, especially in more recent writing (Eisenberg 1994, Batson 1991). The positing of the ego and empathy dynamic goes back to the historical juxtaposition of self-interest or egoism and sympathy or fellow feeling in the thought of David Hume, Adam Smith, and Schopenhauer (Wispe 1991). The present articulation goes back to my doctoral dissertation done at Stanford University (1974). The conflict systems behavioral model was applied in several programs which I authored for corporate management training through the education and consulting corporation United States Education Systems during the period 1976-85. Recently, Roger Masters (1989) has also noted the possible innate roots of contradictory impulses to include selfishness and cooperative or altruistic behavior in human nature. Trudi Miller (1993) has also drawn our attention to this historical duality and suggested its applicability for today. Neither Hume, Smith, Schopenhauer, Wispe, Masters, nor Miller, however, attempted to articulate a model of behavior based upon this duality, or as MacLean calls it, "triality", acknowledging the role of the neocortex in articulating the otherwise nonverbal urgings (1993).

[16]The frontal neocortex especially has long been recognized to be involved in executive functions. See the excellent summary and discussion of findings in Fuster (1997: 150-84). See also Pribram(1973, 1994). Although executive function is frequently equated with frontal cortex function Eslinger(1996) reminds us that the neural substrate of executive functions is better conceptualized as a neural network which includes the synchronized activity of multiple regions, cortical and subcortical (1996: 392). Eslinger also notes the usual neglect of critically important affectively based empathy as well as social and interpersonal behaviors in neuropsychological, information-processing, and behavioral approaches (390-91).

EXECUTIVE PROGRAMMING, NEURAL NETWORKS, AND NEURAL GLOBAL WORKSPACE

Bernard Baars, of the Wright Institute and his colleagues have proposed a Neural Global Workspace Model (GW) which combines the concepts of attention, working memory, and executive function into a theatre metaphor. Baars and colleagues (Newman, et al. 1997; cf. Harth 1997) review other neuroscience and neural network models that deal with attention, binding, resource allocation, and gating that share significant features with their own GW model for conscious attention (for an alternative model based on an evolutionary and clinical approaches and which draws upon MacLean's triune concept, see Mirsky 1996).[17] The authors acknowledge that the models they present implement only partial aspects of their GW theory. Notably neglected are the influences of memory and affective systems upon the stream of consciousness (1997: 1205). Other cognitive metaphors, compatible with GW theory, like Minsky's "Society Theory"(1979) and Gazzaniga's "Social Brain" (1985), remain cognitive in their treatment of sociality although they may be taken to imply affective mechanisms. The CSN model presented in this chapter attempts to incorporate the affective (generalized into empathy) neural substrate necessary to initiate and maintain sociality.

It is noteworthy that 'distributed artificial intelligence' (DIA) models more closely approximate interpersonal behavior in that they seem to reflect an effort at intelligent balance between the competitive self-interest and cooperation which is necessary to the operation of complex social organizations (Newman, et al. 1997: 1196; Durfee 1993). Underpinning the CSN model, the neural substrate for self-survival (generalized as ego) mechanisms may proceed from circuits in the basal ganglia and brainstem (protoreptilian complex) through connections with the amygdala and other limbic structures (early mammalian complex) which add emotion or passion (see Kandel, et al. 595-612), ultimately to be gated into the frontal cortex by thalamocortical circuitry (e.g., see LeBerge 1995; Crick 1994; Baars 1997, 1988).

Likewise, the mammalian nurturing (affectional) substrate and its associated motivation, a fundamental component underlying empathy, may originate in the septal and medial preoptic limbic (see Fleming, et al. 1996; Numan 1994) areas, proceed through hippocampal and other limbic structures, in turn, be gated into the frontal cortex by neuromodulating thalamocortical circuits (to include the cingulate cortex), where the conflict with egoistic imputs is resolved in the executive or Global Workspace of conscious self-awareness.

The neuromodulating and gating of affect, as well as cognition, by the thalamocortical circuitry is supported by neurologists Devinsky and Luciano (1993) who report that the limbic cingulate cortex, a cortical structure closely associated with the limbic thalamus, can be seen as both an amplifier and a filter, which joins affect and intellect, interconnecting the emotional and cognitive components of the mind (1993: 549). Tucker, Luu, and Pribram (1995) speculate that the network architecture of the frontal lobes reflects dual limbic origins of the frontal cortex.

[17]Levine(1986) has also considered MacLean's triune modular concept as a useful tool in network modeling.

They suggest that the two limbic-cortical pathways apply different motivational biases to direct the frontal lobe representation of working memory. They further suggest that the dorsal limbic mechanisms which project through the cingulate gyrus may be influenced by ***social attachments***, and that such projections may initiate a mode of motor control that is holistic and impulsive. In contrast, they speculate that the ventral limbic pathway from the amygdala to the orbital frontal cortex may implement a tight, restricted mode of motor control reflecting the adaptive constraints of ***self-preservation*** (1995: 233-34). These speculative findings are consistent with the CSN model in which ego and empathy represent conflicting subcortical inputs into the cortical executive.

Chapter 5

The Algorithmic Rules of Reciprocal Behavior

The two master, inclusive and modular programs of self-preservation and affection that have been wired into our brain structure operate dynamically according to a set of subjectively experienced and objectively expressed behavioral rules, procedures, or algorithms. Understanding the workings and applications of these algorithms is the key to grasping the role of dialectical conflict and stress in our personal, as well as social, economic, and political lives.

The major ranges of the conflict systems neurobehavioral behavioral model (figure 3) illustrate the features of this ego-empathy dynamic. In the display, subjectively experienced internal as well as interpersonal behavior is divided from right to left into three main ranges called the egoistic range, the dynamic balance range, and the empathetic range. Each range represents a particular mix of egoistically and empathetically motivated behaviors. The solid line stands for ego and pivots on the word "ego" in the executive program of our brain diagram. The broken line stands for empathy and pivots on the word "empathy" in the diagram.[18]

[18]The dynamic of the model, the tug and pull of ego and empathy, self-and other-interest, allows the expression of the mix of motive and behavior as a range or spectrum. The usual dichotomizing of self-interest and altruism is seen only at the extremes of the ranges. All or most of behavior is a mix of varying proportions. Jencks (1990: 53-54) also notes that every motive or act falls somewhere on a spectrum or range between the extremes of selfishness and unselfishness. Teske (1997) sees a blend of self- and other-interest is his identity construction concept. The CSN model moves to identify and explicate some fundamental brain algorithms that provide framework, structure, and dynamic to our socio-experiential performance. At this point it is perhaps also clarifying to acknowledge that the neural mechanisms underlying social behavior may vary among unrelated species to the extent of being entirely different when we move, for instance, from the relatively simple neurological structures of social insects, who apparently function like automatons, to the enormous complexity of the human brain which functions on the basis of choice among conflicting alternatives. That different mechanisms may produce similar results is illustrated dramatically by the evolutionary case of the eye. The eye's evolution was not a process of unfolding developmentally, but rather it developed perhaps 40 different times in evolutionary history, based on at least three functional principles (see Corning 1995: 92-93; also Land and Fernald 1992).

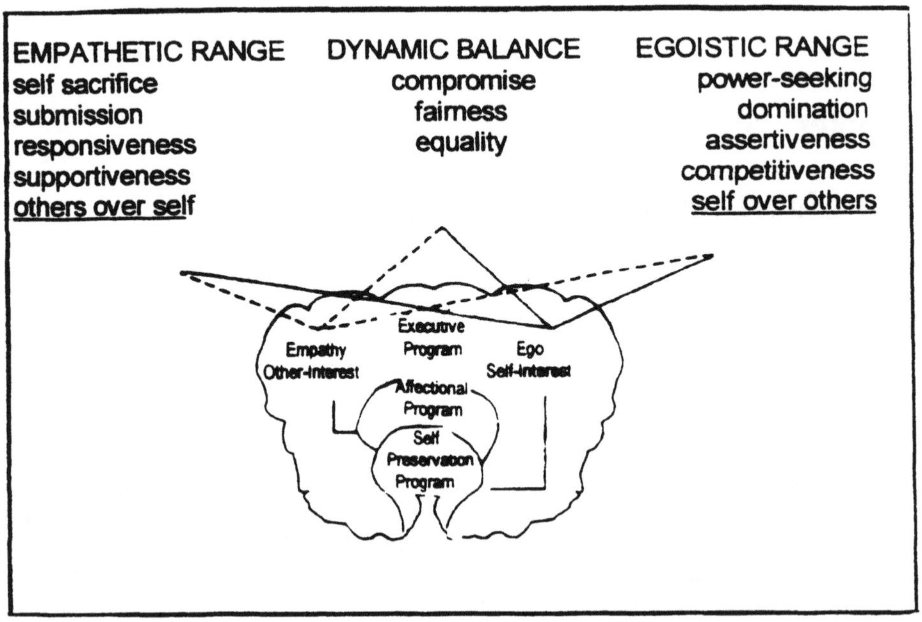

Figure 3. The major ranges/modes of behavior.

The Egoistic Range.

The egoistic range indicates behavior dominated by self-preservation programming. Since the two behavioral programs are locked in inseparable unity, empathy is present here, but to a lesser degree. Behavior in this range is self-centered or self- interested and may tend, for example, to be dominating, power-seeking, or even attacking, where empathy is less. When empathy is increased, ego behavior will become less harsh and may be described more moderately as controlling, competitive, or assertive. As empathy is gradually increased, the intersection of the two lines of the diagram will be drawn toward the range of dynamic balance. Ego behavior will be softened as empathy is added. But the defining characteristic of the egoistic, self-interested range is self-over-others. Whether we are blatantly power-seeking or more moderately assertive, in this range we are putting ourselves, our own priorities, objectives, and feelings, ahead of others. We're telling others, "me first."

The Empathetic Range.

The empathetic range represents behavior weighted in favor of empathy. Ego is present, but is taking a back seat. When ego is present to a minimal degree, empathetic behavior may tend to extremes of self-sacrifice and submission. When ego is increased, empathetic behaviors become moderated and may be described as supportive, responsive, or any of a variety of "others first" behaviors. As the influence of ego is gradually added, empathetic behavior will approach the range of dynamic balance. In the empathetic range, the key phrase to remember is others-over-self or others first. Whether we are at the extreme of self-sacrifice or more moderately responsive, we are putting the priorities of others ahead of our own.

The Dynamic Balance Range.

The range of dynamic balance represents a working balance between ego and empathy. At this point our behavioral programs are operating in roughly equal measure. I speak of "working," "rough," or "dynamic" balance because the tug-and-pull between the two programs continues ceaselessly. The dynamic nature of the programming means that "perfect" balance may be a theoretical point, unattainable in practice. Our more balanced behavior tends to be characterized by equality, justice, sharing, and other behaviors that show respect for ourselves and others. In fact, respect for self and others is the keynote of the range of dynamic balance.[19] We are working to achieve shared priorities, objectives, and feelings.

Energy or Activity Level.

The extent to which the master modules or programs of self-preservation and affection, ego and empathy, are out of balance, or pulling against each other, is a measure of behavioral tension. We experience this behavioral tension both internally and between ourselves and others, in any relationship or interaction. Unmanaged or excessive tension becomes, of course, behavioral stress. But that's not all. Important also is the degree of energy we give to the interaction or the relationship. The amount of energy we put into any activity depends mostly upon how important we think it is or how enthusiastic we feel about it. In competitive sports or contests, qualitative differences in energy are easily observed. In intellectual contests, like chess, the energy may be intense, but less obvious. Although self-preservation and affectional programs -- ego and empathy -- are the main variables of interpersonal behavior, energy or activity level is often important to consider in describing more fully our interactions.

THE PROPOSED OPERATING ALGORITHMS OF INTERPERSONAL BEHAVIOR

From the dynamic interplay of ego, empathy, and activity level come the following rule statements.

1) Self-interested, egoistic behavior, because it lacks empathy to some degree, creates tension within ourselves and between ourselves and others. The tension increases from low to high activity levels. And it increases as we move toward the extremes of ego.

Within ourselves, the tension created by the tug of neglected empathy is experienced as a feeling of obligation to others or an expectation that they might wish to "even the score" with us.

Within others, the tension created by our self-interested behavior is experienced as a feeling of imposition or hurt, accompanied by an urge to "even the score."

Children often reveal the dynamic of such behavior in a clear, unsophisticated form. Imagine two children playing on the living-room floor. One hits the other.

[19]See Eckel and Grossman (1997). Without making any connection with brain science or the reciprocal laws of behavior, the authors use a typology of fairness (for me, for you, for us) which expresses the conflict systems model and the reciprocal algorithms of behavior.

The second child hits back, responding in kind. Or the children may not hit each other at all. One might instead call the other a bad name. The second child reciprocates, kicking off a round of escalating name-calling. One child may eventually feel unable to even the score and will complain to a parent to intervene. Most of us have experienced such give-and-take as children and have seen it countless times in our own children and grandchildren. Similar behavior is embarrassingly observable among adults. It can be seen in husband- and-wife arguments, bar fights, hockey games, political campaigns, even in sophisticated lawsuits. The rule operates not only in such highly visible conflict situations, but also in very subtle interactions -- in the small behavioral exchanges, the ongoing give-and-take of all interpersonal relations.

By analogy with Newton's third law of motion, which states that for every action there is an equal and opposite reaction, we can say that

the reactions that build in ourselves and others do so potentially in proportion to the behavioral tension created by the egoistic, self- interested behavior.

That is, the harder I hit you, the harder you hit me in return. Or the fouler name you call me, the fouler name I call you in return. Or perhaps with more sophistication, I resolve the tension in me by an act of visible "superiority." I ignore you -- although I could call you an even fouler name, if I chose.

Behavior on the other side of the scale is described in the second rule statement:

2) ***Empathetic behavior, because it denies ego or self-interest to some degree, also creates tension within ourselves and others. This tension, likewise, increases as activity levels increase and as we move toward extremes of empathy.***

Within ourselves, the tension created by the tug of neglected self-interest (ego) is experienced as a feeling that "others owe us one" and a growing need to "collect our due." This tension, especially if it continues over time, may be experienced as resentment at being exploited, taken for granted, not appreciated, or victimized by others.

Within others, the tension created is experienced as a sense of obligation toward us.

The reactions that build in ourselves and others, again, are in proportion to the behavioral tension created. And again, the unmanaged, or excessive tension is experienced as behavioral stress.

When we do things for others -- give them things, support them, make personal sacrifices for them -- it can make us feel good, righteous, affectionate, loving. But when we do these things, we do want a payback. That's the tug of self-interest. It can be very slight, hardly noticeable at first. But let the giving, the self-sacrifice, go on for a while, unacknowledged or unappreciated (that is, without payback to ego), and see how we begin to feel. The tension, the stress, starts to show. We complain that others are taking advantage of us, failing to appreciate us, taking us for granted, victimizing us. Self-interest cannot be long short- changed without demanding its due. We may eventually relieve the stress by blowing up at those we have been serving -- accusing them of ingratitude, withdrawing our favors, or kicking them out of the house. Or we may sandbag the stress, letting it eat away at our dispositions, our bodies.

On the other hand, when we do things for others, they often feel obliged to return the favor in some form to avoid being left with an uneasy sense of debt. Gift-giving notoriously stimulates the receiver to feel the need to reciprocate. Think of the times when you have received a holiday gift from someone for whom you had failed to buy a gift. Sometimes the sense of obligation prompted by the empathetic acts of others can become a nuisance.

The third rule statement describes the balance between these two extremes:

3) ***Behavior in the range of dynamic balance represents an approximate balance of ego and empathy. Within ourselves and others, it creates feelings of mutuality and shared respect.***

For most of us it's an especially satisfying experience to interact with others in equality, with no sense of obligation, superiority or inferiority. To work together in common humanity, in common cause, is to experience behavioral dynamic balance. Of course, there are many versions of the experience of dynamic balance: the shared pride of parents in helping their child achieve, the joy of athletes in playing well as a team, the satisfaction of co-workers in working together successfully on an important project.

The Reciprocal Nature of Behavior.

These algorithms of behavior operate in the smallest interactions, the vignettes, of everyday personal life. The dynamic of behavioral tension dictates that for every interpersonal act, there is a balancing reciprocal. A self-interested act requires an empathetic reciprocal for balance. An empathetic act, likewise, requires a balancing self- interested reciprocal. This reciprocity goes back and forth many times even in a short conversation. Without the reciprocal, tension builds, stress accumulates, and either confrontation or withdrawal results.

Reciprocity through Conflict.

These, then, are the proposed basic interpersonal algorithms of our three-level brain. These algorithms show how we get to reciprocity through conflict. I propose that they shape the conflict and reciprocity, the give- and-take, at all levels of our interactive, social lives.

Overemphasis on either self-interest or empathy, exercise of one program to the exclusion of the other, creates tension and stress in any social configuration -- from simple dyadic person-to- person encounters up to and including interactions among members of the workplace, society at large, social groups, and entire economic and political systems.[20]

[20]Somit and Peterson (1997) see that evolution has provided us with a predisposition for hierarchically structured social and political systems, in other words a tendency to hierarchy. I would suggest that this may be seen as an alternative perspective of the same dynamic of the tug and pull of ego and empathy of the triune brain structure that underlies reciprocity. Reciprocity, although more often than not seemingly unbalanced, in social and political relationships, is nevertheless always there to some degree. Even the range of dynamic balance of the conflict systems model is an approximate and shifting balance with some degree of hierarchy of dominance and submission. In its purest form, ceteris paribus, the innate dynamic only tends, rather imperfectly toward a balanced reciprocity. It does not and cannot achieve it deterministically, but only probabilistically. When other things are not equal, i.e. there are differences in personal strength, talent, ability, and intelligence, such differences will allow some individuals or groups to control more resources and thereby

THE QUESTION OF SCIENCE: PHYSICS VS. SOCIAL

The proposed reciprocal algorithms of behavior can be viewed as high-level brain algorithms, built up from a nested hierarchy of interconnected lower-level algorithmic modules.[21] The algorithmic rules, as proposed here, operate very imperfectly. I suspect that this will be true of any behavioral algorithms or principles, proposed at this level of generalization. The proposed algorithms, then, can approximate, but not fully achieve, the precision of the laws of classical physics or even quantum mechanics. This is because they are achieved through the process of organic evolution (which involves some chaotic and random processes as well as natural selection) and therefore cannot operate as immutable universal physical laws but as generalized algorithms with degrees of variation.

The idealized, or rather statistically generalized, tug and pull of ego and empathy presented here may be further probabilized in actuality by genetic, gender and developmental, individual experience and learning, and other environmental shaping and reinforcing factors.

In other words, genetically speaking, given the individual differences in genetic inheritance that we see in such obvious things as in hair, skin, or eye color, some individuals behaviorally may be more or less as strongly wired for self-preservation and affection as others. And granting gender and developmental differences, every human being is, nevertheless, similarly wired with the fundamental brain architecture unless he/she has very serious genetic defects indeed. Influential developmental psychologists like Jean Piaget (1965) of Switzerland and Lawrence Kohlberg(1984) of Harvard, operating from a behavioral perspective, have constructed and tested theories of childhood moral development. In the theories of both men moral stages of development emerge much the same in all cultures when

---

create hierarchies. Such hierarchies may be accepted by the less capable, but not without behavioral tension. The hierarchies are inherently unstable because of this behavioral tension. That is, as soon as the unequal capacities become less unequal, those on the lower end almost invariably move to contest and alter the hierarchy. Somit and Peterson acknowledge this tendency to unbalanced reciprocity or hierarchy as regrettable. And they devote their book to helping us understand how we may achieve the desired, balanced or non-hierarchical political system. Salter (1995) has pulled together a considerable quantity of naturalistic observation on the genetically-based communicative signaling in human and nonhuman species involved in the negotiating and maintenance of hierarchies.

[21]The algorithms of reciprocal behavior may also be thought of as high level Darwinian algorithms (see Cosmides and Tooby 1989 and Tooby and Cosmides 1989), which function as the cognitively generalized sum of perhaps many contributing and perhaps more highly specific innate algorithms (see also, Cory 1996; cf. Vandervert 1997). It might be further noted that the CSN model, which rests upon evolved algorithms of the brain, may be consistent with the sensory motor approach to cognition (e.g., see Newton 1996) as long as the very extensive and complex algorithmic processing and structuring of sensory and motor inputs and outputs is not treated too simplistically. For example, Harth comments that the massive loops of reciprocal connectivity between the cortex and the subcortical relays in the visual system give the impression that "the cortex is more bent on introspection and confabulation than forming an unbiased view of the outside world"(1997: 1245).

the child experiences anything approaching a normal family life. Such generalized moral stages could not be found across cultures if they were not genetically based on the species-wide brain structure and its associated behavioral potentialities.

From the standpoint of individual learning, socialization, and other environmental factors, modifications in biological structures and potentialities occur in early development and throughout life. Individual life experiences may facilitate, suppress, strengthen, or otherwise channel, the expression of these inherited biological programs. Environmental factors, to include physical constraints as well as our socially and scientifically accepted institutions and paradigms, may also shape and reinforce the expression of the evolved algorithmic dynamic. Individual learning, experience, or environmental factors of the individual life cannot, however, eliminate the genetic structure and programming of the brain...that is, not without radical injury, surgical or genetic intervention. And the behavioral tension will be there to both resist the changes and to shape the experience, even shape the environment itself, in a dynamic manner.

Because of these factors, the behavioral algorithms are statistical ...much in the same way as are the second law of thermodynamics and quantum theory of physics. That is, they do not allow precise prediction of specific behavior at the basic unit of analysis...the individual, molecular, or subatomic level respectively...but only on the aggregated basis of statistical probability. The proposed algorithmic rules of reciprocal behavior, as here presented, may nevertheless very well prove to be equally as valid and useful to social science as the laws of physics are to physical science. They do not and cannot, however, have the immutable quality of physical laws such as gravity. As products of organic evolution, they inevitably involve more probabilities because of individual differences, genetic and learned, in the evolved basic units.

It is also interesting to note that an analogy can be made between the inclusive spectrum of possible behaviors of the conflict systems neurobehavioral model and the particle/wave function of quantum physics. As the wave function of a particle is defined to include all the possible values of a particle according to probability, the wave function of behavior can be thought to include all possible internal and interpersonal behavioral probabilities (mixes of ego and empathy) extending across the egoistic, empathetic, and dynamic balance ranges. Externally, observed behavior is predictable from the model as is quantum behavior, only on a probability basis specified by the wave function. The behavioral wave function, like that of particle physics, collapses or reduces to one behavior in a decision, action, or observation. If it doesn't collapse, we see frustration, tension, and indecisiveness... ambiguous behavior stalled in uncollapsed waveform.

Upon observation by an external observer, the wave function of behavior can be considered to collapse to a specifically observable behavior on the part of the individual and that's the end of it. But internally, we experience it differently because we have access to the dynamic. We know, in our conscious awareness, the tension, the difficulty, the struggle we go through in important issues of ego and empathy conflict. In physics, however, we, perhaps, simply do not yet understand what set of dynamics leads to the wave function collapse.[22] In behavior, the dynamic lies in the complexities of subjective preconsciousness and/or self-aware consciousness.

[22]That is, in physics it is not known exactly why and how wave function collapses or reduction occurs and how eigenstates are determined (e.g., see Hameroff and Penrose 1996: 311). The standard Copenhagen Interpretation sees collapse as occurring at randomly measured values when the quantum system interacted with its environment, was otherwise measured, or consciously observed; (e.g., see Stapp's 1972 well-known article on the Copenhagen Interpretation). Penrose (1994) and Hameroff and Penrose (1996) introduce a new physical ingredient they call objective reduction (OR) in which quantum systems can self-collapse by reaching a threshold related to quantum gravity. Ellis has compared consciousness to a wave pattern or function (1986: 67). Harth notes, in summarizing his sketchpad model, that "the transformation from the extended activities in the association areas and working memory to specific mental images may be likened to the collapse of a wave function in quantum mechanics..." He does not, however, imply any quantum effect(1997: 1250).

Chapter 6

The Conflict Systems Neurobehavioral Model vs. the Maslow Hierarchy

It is useful at this point to return to the Maslow need hierarchy and contrast it with the conflict systems neurobehavioral model developed by building upon the triune brain model of MacLean.

Maslow's focus on a staircase-like hierarchy of inner needs tended to turn us inwardly away from the social environment. Co-opted and blended with the prevailing view of self-interest as the dominant human motive, Maslow's theory of self-actualization, with its lofty connotations of self-fulfillment and creative expression became reduced, especially in the decades of the 70s and 80s, to a license for indulgent self-interest (Yankelovich 1981; Cory 1992). This isolated, indulgent version of self-interest, as expressed in our social and business experience, earned the labels of "narcissistic" and "me first." One of the great popular appeals of the Maslow hierarchy is that it fit so well with the prevailing emphasis on self-interest in our everyday as well as academic thinking on economics and politics.

CONFLICT, NOT EMERGENCE.

But inward focus and simplistic hierarchy were not the only problems with Maslow's hierarchy. It allowed us to be drawn excessively toward our inner selves and away from society because it missed a point central to human behavior--that conflict, not emergence, is behavior's most definitive characteristic.

Maslow placed the social or relatedness needs (empathy) lower on the escalator than esteem and self-actualization needs (ego). His theory contains the clear suggestion that we pass through these social needs or rise above them in the trek up the hierarchical ladder. This was a fundamental error, resulting in a considerable distortion of our view of human social nature. The two sets of needs, social and self-interested, although hierarchical to some extent, are wired together in the same brain to produce the tug and pull, conflictual dynamic of interpersonal or social behavior. We are not autonomous, but at best semi-autonomous creatures, completely immersed in a pervasive social context, which is, at this point in our evolution, both demanded and made possible by our evolved brain structure.

FOR ME? FOR YOU? OR FOR US?
IT'S ALWAYS THE SAME!

The tug and pull of ego and empathy drives interpersonal behavior not only at the levels of Maslow's social and esteem needs, but at all levels. The basic questions of interaction are the same anywhere in the hierarchy.

These questions are: Do I do it or take it for myself (ego)? Do I do it for or give it to others (empathy)? Or do I do it for both myself and others? (dynamic balance).

It is enlightening to try out these questions at each level of Maslow's staircase. Take the physiological needs first. I can see myself lost in the desert with a friend. I have one remaining canteen of water. It's half full, and I don't know if or when we'll find more. My interpersonal choices are three: Do I keep it all for myself (ego)? Do I give it all to my friend (empathy)? Or do we share it (dynamic balance)? I would face the same choices if my companion were a spouse, child, friend, or enemy.

These three questions would likewise apply to the next level of needs, those of safety. Suppose we are threatened by a wild beast? Or a natural disaster? An intruder? A terrorist? Do I protect myself? Others? Self and others? The conflict is more or less evident depending on the urgency of what is happening, but it is always present.

At the next level the social needs are directly related to affection and empathy, while above them, the esteem needs are related to ego. But as we have seen ego and empathy and their inexorable conflict pervade all three levels of Maslow's hierarchy. In evolutionary terms, the conflict did not exist prior to the appearance of affectional programming in the mammalian brain; but ever since it appeared, it has influenced all needs and all behaviors.

The ego-empathy conflict even pervades Maslow's highest need level, self-actualization. Behaviorally speaking, the need for self-actualization has to do with ego rather than empathy, but again the choices are the same as at all the other need levels. Do I put my own priorities, feelings, and objectives first? Or do I first consider the priorities, feelings, and objectives of my parents, spouse, children, friends, company, church, nation, or world-wide humanity itself? Or do I struggle to achieve a balance?

Chapter 7

The Reciprocal Algorithms of Behavior and the Norm of Reciprocity

The norm of reciprocity has been a major theme in anthropology and sociology for the better part of a century. This universally observed norm has been accounted for, or shown to be possible, in evolutionary theory by such concepts as kin selection, inclusive fitness (Hamilton 1964), reciprocal altruism (Trivers 1971, 1981; Alexander 1987), and game theory (Maynard Smith 1982, Axelrod and Hamilton 1981, Bendor and Swistak 1997). These efforts draw upon gene-centered perspectives, which see such reciprocity as basically selfish. More recently extensive reciprocity seen as based not upon selfishness, but empathy has been reportedly observed in the behavior of rhesus monkeys (de Waal 1996). F. de Waal's approach is a welcome departure that tries to escape the selfishness of gene-centered approaches and looks to the implied motivational mechanisms. All these approaches, however, to include that of de Waal's, have been based on the external observation of behavior. They have not attempted to identify or even speculate upon the neural mechanisms within the organism that must necessarily have been selected for by the evolutionary process to accomplish the functions of motivating, maintaining, and rewarding such observed reciprocal behavior.

I suggest it is time now to shift emphasis from the old debates from the gene-centered and ethological perspectives to the actual mechanisms of the human brain. I think that at this time in our evolutionary thinking, it has been established beyond any reasonable doubt by the work of such researchers as Hamilton, Trivers, Alexander, Maynard Smith, et al., that even from the most hard core selfish gene perspective, the basis for the closely related behaviors of reciprocity, cooperation, and altruism has, from the Darwinian or Neo-Darwinian perspective, been established in the human genome (e.g., see Corning 1996 for a summary supporting this argument). The presence of these behaviors has further been confirmed by quantities of observational data in primates, even in studies of early protohuman hominids (Isaac 1978), and by extensive anthropological and sociological observation.

In other words, we now know that we must have, wired into our brain and nervous system, the neural mechanisms that make such behaviors possible. It is time, therefore, with the full emergence of neuroscience, to make every effort to identify and specify these brain mechanisms and extrapolate the implications of their presence and functioning for our social and political lives. This is, in fact, the thrust of the emerging subdiscipline of evolutionary psychology (Cosmides and Tooby 1989, Tooby and Cosmides 1989, Barkow, Cosmides, and Tooby 1992).

Understanding the neural reward systems and the internal dynamic of the triune modular brain structure in humans is critical to properly understanding our human social life because the dynamic of such reward systems provides the subjective motivational basis for our choices in behavior as well as the entire texture and meaning of our lives. This subjective motivation and experience, although it is the most important aspect of our lives, has, of course, been almost completely ignored by the externalized gene-centered perspective. It is time to acknowledge more fully this subjective motivation along with its objective manifestations and give it its due place in our lives. The proposed algorithms of reciprocal behavior based upon the tug and pull of ego and empathy of the triune brain structure help explain much of what has been observed and reported over the last few decades. A brief overview may help to make this clear.

Over 35 years have passed since the publication of Alvin Gouldner's (1960) much referenced article on the norm of reciprocity in the *American Sociological Review.* Gouldner cited much of the literature in anthropology and sociology existing at that time. L.T. Hobhouse writing at the beginning of the century called reciprocity "the vital principle of society"(1906:12). Writing somewhat later Richard Thurnwald described it as a principle "which pervades every relation of primitive life"(1932:106). Leading scholars, such as anthropologist Bronislaw Malinowski (1922, 1926) and sociologist Georg Simmel (1950:387) have considered it the sine qua non of society. Harvard sociologist Talcott Parsons viewed it as "inherent in the nature of social interaction"(1951:21). French anthropologist Claude Levi-Strauss has referred to it as a "trend of mind" (1969:98). More recently J. van Baal (1975:12) has considered it a "part of the human condition."

So universally has reciprocity been observed that leading anthropologists and sociologists have long suspected that it must have a psychological or biological origin (e.g., see Levi-Strauss 1969 and Homans 1950; 1961:317). That is, reciprocity must ultimately rest upon mechanisms within the human brain. The proposed algorithms of reciprocal behavior deriving from the dynamic of our evolved modular brain structure provide these mechanisms. A closer look at some of the dynamics of exchange will bear out the empirically observable workings of these algorithmic processes.

Gift-giving in Primitive Societies.

French anthropologist, Marcel Mauss, wrote the classical study on gift-giving in primitive societies. His study, called *The Gift* (1954), covered the potlatch system of the Northwest American Indians. Although exaggerated in its rituals of display and presentation, the potlatch tradition follows the algorithms of reciprocal behavior.

According to the reciprocal algorithms gift-giving notoriously stimulates the receiver to feel the need to reciprocate. A disruption of reciprocity creates behavioral tension, stress accumulates, and if reciprocity is not restored either confrontation or withdrawal results. In the potlatch system failure to reciprocate adequately resulted in loss of social status. In the potlatch system, as well as in other primitive societies, a refusal to accept or reciprocate gifts may be tantamount to a declaration of hostilities (Mauss 1954: 79). This balancing act of reciprocity is not only operative in the daily interactions of our modern lives, but pervaded the social and exchange activities of primitive societies. J. van Baal observed that unbalanced and balanced reciprocity seemed directly connected with inequality and equality in primitive societies (1975:29).

Anthropologist, Marshall Sahlins, at the close of his comprehensive article on the sociology of primitive exchange found that reciprocal interaction pervaded all levels of social life in primitive societies. Sahlins observed that "...Here has been given a discourse on economics in which 'economizing' appears mainly as an exogenous factor!" Sahlins went on to point out that since, in his study, the organizing principles of economy have been sought in social factors other than the self-interested, hedonistic tendency, his study suggests a strategy that is the reverse of economic orthodoxy (1963:186). And, in a later work, discussing a properly anthropological economics, he writes that "...every exchange, as it embodies some coefficient of sociability, cannot be understood in its material terms apart from its social terms" (1972:183).[23]

Studies of social exchange in primitive societies, then, tend to show that exchange rests not primarily on what are assumed to be economic, but rather on social factors. This reflects the algorithms of reciprocal behavior. The overriding fact is that humans socialize. They are mammals and the social glue emerges from the bonding mechanisms of the mammalian affectional programming. These bonding mechanisms potentially conflict with the earlier protoreptilian programming from which ego and self-interest emerge. The neocortex evolved with its capacity for generalization and language and the tug and pull...the conflict of ego and empathy, self and other interest was on! The reciprocal algorithms of behavior emerge from this triune neurological configuration.

In modern economics with its almost exclusive emphasis on self-interest, it is sometimes forgotten that human actions are invariably within a social context...even withdrawal and hostility have a social context. Self-interest, moreover, is usually if not always, necessarily defined in social terms and is completely immersed in the social context. Modern capitalist economics, particularly, has exaggerated the singular pervasiveness of self-interest. It has done so by focusing exclusively upon self-interest and assuming away the pervasive social context[24] in which it operates reciprocally with empathy. A brief survey of the evolution of the market will bear this out.

[23]For a study of reciprocity from the perspective of economics, which uses game theory and draws upon anthropology and evolutionary psychology, see Hoffman, McCabe, and Smith (1998). The researchers found more trust and stable reciprocity than anticipated.

[24]Many scholars have criticized economics for ignoring the social context. For example see Gill (1996, esp. pp. 152-53).

Chapter 8

Empathy in Economics: Anthropological and Sociological Perspectives

To understand the behavior of the modern day free enterprise market as it is shaped by our inherited brain structure and behavior, it is helpful to go back to early times--to reconstruct as best we can the days before the market appeared.

THE FAMILY OR GROUP BOND.

In those times, when people consumed what they produced, the excess which they shared with, gave to, or provided for family or community was in the nature of natural affection or empathy. The reward for the empathetic, providing act was emotional --there was not a specific, but a diffuse value assigned to it. It also had social effects -- the givers, providers gaining status in the group. The emotional and the social effects were both directly governed by the reciprocal algorithms of behavior.

Let's look more closely. The provider, say the warrior brought meat from the hunt or the wife brought berries and fruits from the field, tanned skins, etc., to give to the family or group (cf. Willhoite 1981: 242). The act of providing, giving, created behavioral tension in the giver, who acting empathetically denied ego to some degree and required a response of acknowledgment, gratitude, respect, affection, or some other reaffirmation of ego. This providing or giving also created behavioral tension in the receivers. It was a service to their ego -- their own preservation -- which created tension requiring an offsetting empathetic response, a thank-you, an expression of appreciation or respect. In any family or close group, even now, this dynamic flows constantly, even in the smallest activities. In the small group the rewards, the reciprocations, are largely not quantified, but are diffuse. They become obligations -- bonds -- which hold the group together for protection or mutual survival. Nevertheless, they must achieve some approximation of balance or the unresolved tension will build within the group and become disruptive.

THE GIFT

From these early, primitive behavioral exchanges, emerged the gift: an empathetic act of providing or serving that followed the same algorithmic behavioral rules that governed provision for survival. It created tension in the giver--an expectation of reciprocity -- and tension in the receiver who was bound to reciprocate. The rewards associated with the gift were diffuse, unspecified, unquantified -- except by some subjective measure of feeling, emotion, or behavioral tension. A gift to a warrior or chief might vaguely obligate his protection. A gift to a prospective mate might vaguely obligate his or her attentions. As the group became larger, beyond immediate kin, perhaps equivalent to a village, some people became noted for their special abilities, talents, or services. In such an early pre-market economy the gift evolved toward a new, modified form -- the gift exchange.

THE GIFT EXCHANGE.

There was a young mother of the village who had a sick child. The child burned with fever and could hold no food. Despite all the mother did, the child seemed only to get worse.

In the village lived an old hag, who was rumored by all to have powers to cure sickness. The very thought of this hag struck terror into the young mother's heart. As a girl she and the other children had jeered and thrown rocks at the old crone plodding the paths of the village. She feared the hag but she feared still more that her child would die. At last she plucked up her courage and carried her child to the hut at the far end of the village.

At the door to the hut, the mother and child were received without words. For many hours the old hag prepared and gave to the child elixirs of herbs and earth. She bathed, wrapped, and warmed the child over and over as the fever ran its course.

The hag worked quietly, deliberately, with expressionless concentration as she ministered to the child. The mother watched in hushed anxiety through endless hours from her seat in the corner of the hut.

At last the fever broke and the child could take food. The young mother was greatly relieved. She left the child in the care of the hag and was gone for several days.

When the young mother returned for her child, she brought a small cloth filled with shells which she laid at the feet of the crone seated in a corner of the hut.

The old hag took the cloth and opened it slowly. As her gnarled fingers rolled thoughtfully over each of the pearly, luminescent, exquisitely formed shells, she knew that the young mother had brought them from the far shore. Her tight, withered jowls quivered slightly as the leathered line of her cracked lips turned up -- slightly at first – then broadly.

And for the first time, the young mother saw not a hag...but an old woman...who smiled.

In this short story we see all the behavioral elements of the gift combined with the beginning of an exchange for services. The act of empathy, of service, of providing, reciprocated by a confirming act of labor and value. Although unspecified at first, the reward or payment for services would probably become standardized through repetition. And it would be accompanied by the effect of obligation, of bonding. The relationship between the old crone and the young mother is changed forever. The young mother no longer sees her benefactor as a frightful old hag, but as a loving old woman.

The true character of the marketplace, then, harks back to the nature of the gift. The producer or provider offers, in empathy, a gift of self--of time, effort, labor, creativity. The buyer or customer acknowledges this gift in receiving it by a return gift, usually in the form of money that represents the accumulated time, effort, work, and creativity of the buyer. It all follows the reciprocal algorithms of behavior.

FROM GIFT TO TRANSACTION.

From the gift evolved the transaction -- namely the gift with the reciprocal specified or quantified. The transaction is the beginning of the contract, perhaps of the market itself. The transaction operates, however, by the same algorithms of behavior as the gift -- except that it attempts to head off the residual, unresolved behavioral tension that creates a condition of obligation or bonding. Nevertheless, it retains its essential mammalian characteristics as an act of empathy, of nurturing, which requires a balancing reciprocal act in payment to ego.

When we encounter its equivalent in today's impersonalized market economy, how often do we feel the subjective experience of the transaction? We take our sick child to the doctor, who empathetically and carefully applies the knowledge that took ten years and a fortune to gain. We pay the bill -- that is, we make a return gift with money which represents a portion of our accumulated education and labor. The scenario is repeated in transactions with the plumber, the carpenter, the computer maker. The behavioral algorithms still apply, but the feeling, the subjective experience has to a large degree been lost.[25]

EMPATHY IN THE MARKETPLACE.

The overemphasis on self-interest and the lack of an adequate behavioral model have prevented us from seeing how the marketplace derives from brain structure, and how empathy or altruism plays an equal role with ego or self- preservation. But the role of empathy is clearly present in the language, if not the practice, of the marketplace. The everyday language of marketing is the language of empathy. Advertisements, almost without fail, emphasize service or benefit to the customer.

Customer service is, in fact, the keynote of most businesses. Every retail store of any size has a customer "service" department in a prominent location. Almost nowhere else are we treated with more exaggerated empathy, even obsequiousness,

[25]The discussion here draws extensively upon Cory (1996). From the functional viewpoint in sociology this shift from gift to transaction and the subsequent expansion of social organization and the market, which involved the depersonalization of ego and empathy, is captured by Parsons in his well-known pattern variables. These variables describe the shift from the predominantly personal and stable relationships of primitive societies to the largely impersonal or business-like relationships of modern society. The relationships shifted from ascription to achievement; diffuseness to specificity; affectivity to neutrality; particularism to universalism; collectivity to individualism (self). See Parsons (1951: 77; also, 1960).

than in some retail businesses. In marketing, as any good salesperson can tell you, empathy works -- not so much because people enjoy expressing it, although they do, but because people respond to it. The features-and-benefits approach to selling is built entirely on empathy. Depending on the skill and sincerity of the salesperson, the genuine feeling may or may not be present, but the basic empathetic behavior is there.

The act of selling thus clearly demonstrates not only the role of empathy, but also the reciprocal algorithms of behavior. The salesperson's serving behavior toward the customer creates tension in the salesperson, who, acting empathetically, denies his or her own ego to some degree, thereby producing the expectation of reciprocity, a reward. The salesperson's empathetic behavior also creates tension in the customer-- a sense of obligation to buy. The sales process not only demonstrates these reciprocal algorithms; but it depends upon them. We see the driving, contending forces of our evolved brain dynamic, at work in the daily activity of selling and buying. Anyone who focuses only on self-interest sees only one side of this behavioral economic equation.

The trick or deception of assigning a self-interest motive to everything -- even to the most empathetic or altruistic acts -- is made plausible by the fact that the reciprocal is always there. There is always an egoistic reciprocal to any empathetic act; and, likewise, there is always an empathetic reciprocal to any egoistic act.

THE DYNAMICS OF POWER.

The ideal marketplace, as well as the ideal personal relationship, assumes a meeting of equals; that is, all participants can reciprocate equally. This is true of primitive as well as modern societies. This ideal is seldom reached in either reality. As noted earlier, status was gained by the givers or providers in a group.

Power, likewise, derives from the differential capability to provide.[26] Those who control a favorable proportion of valued resources or services have potential, if not actual, power over others to the extent of the imbalance. Absolute control of resources and services equates to absolute power. In exaggerated form, this was the effect of the potlatch system of the North American Indians...to create and support status and power. The less powerful depend to some degree on the empathetic behavior, the supplying or provisioning of the powerful, to meet their survival needs. The recipients, since they can't reciprocate equally, become bound by the excess of behavioral tension to an empathetic stance of submission and obedience to the providers. In keeping with the algorithms of reciprocal behavior, such relationships are always accompanied by behavioral tension. And this dynamic of behavioral tension drives the efforts of all parties to alter the balance in their own favor.

[26]Not only Blau (1964), but also Emerson (1972a, b), Cook (1987, 1990), Blood and Wolfe (1960), also, from one perspective or another, see the exchange relationship at the core of inequality.

Chapter 9

Rational Choice Theory Contra the Human Mammal

Rational choice and exchange theory in sociology deal with the issues of power and social integration. Sociologist Peter Blau (1964) hits at the heart of reciprocal exchange when he sees its two general functions as creating bonds of friendship and establishing relationships of subordination or domination. Of course, domination and subordination carry inevitable behavioral tension which is at the root of the underlying implicit assumption of all exchange theory ...the tug and pull tendency toward equalization of value among all social relationships.

The recent work of sociologist James S. Coleman illustrates the distortion, however, that appears in the work of some of our most capable theorists when our mammalian heritage, is not fully grasped. Coleman bases his theory of action on the overt premise of rational, self- interested humans, unconstrained by morality, norms, or altruism. He apparently is unaware of, or else assumes erroneously that we lack, our limbic septal (to include our medial pre-optic area) and thalamocingulate brain structures and that we are only wired innately for self- interest when he writes:

> ...To assume that persons come equipped with a moral code would exclude all processes of socialization from theoretical examination. And to assume altruism or unselfishness would prevent the construction of theory about how persons come to act on behalf of others ...when it goes against their private interests (1990: 31-32).

The above quoted statement is, of course, not true. If we recognize that the programming for ego and empathy, self and other interest, although admittedly undefined by specific social content, is innate in our mammalian neurological structure, we are certainly not thereby prevented from constructing theories about socialization. The focus of such theorizing would, of course, shift from the one-sided view of exclusively self-interested, half-humans to a more correctly balanced view. Our theory could then appropriately focus on how these two innate potentialities of ego and empathy achieve social expression and how they are blended in balanced or unbalanced reciprocity over history in the particular moral codes and norms of any specific culture or society.

Exchange theorist Karen Cook and her associates appropriately see exchange as the ubiquitous structuring activity of societies. Drawing on a metaphor of James

Coleman, Cook writes that theories of social structure alone "provide a chassis but no engine." Concerning the unique contribution of exchange theory, she goes on to claim:

> A major strength of exchange theory is that in making explicit the reciprocal nature of social interaction it provides a theoretical engine of action to power the chassis that is our understanding of social structure (Cook, O'Brien, and Kollock 1990: 164).

Cook correctly sees the reciprocal nature of social interaction as the dynamic process of social action and society-building. She remains, however, within the rational choice, self- interested framework and thereby misses the essence of reciprocity ... the tug and pull of ego and empathy, concern for ourselves, and recognition and concern for the interests of others. Her "engine", being universal, inevitably implies a grounding in biology or human nature. She does not, however, provide such a grounding.

EQUITY THEORY

Equity theory, a recent direction in sociology and social psychology,[27] is closely akin to exchange theory. Equity theory purports to be a general theory that provides insights into social interactions of all kinds...from industrial relations to issues of justice in more general social encounters. Walster, Walster, and Berscheid (1978) provide a thorough statement of equity theory and bring together the previously scattered research as of the year of publication. More recently, it has been extended to the most personal relationships of marriage and other relationships of intimacy. Hatfield and Traupmann (1981:165-178) summarize the application of equity theory to intimate relationships, a focus in social psychology that has emerged in the last two decades.

Equity theory, by the very use of the term equity, reveals its implicit grounding in an intuitive perception of our triune brain function as it tends toward balance in our behavior. Equity theory's four propositions, proclaimed apparently from faulty intuition or generalized observation, are otherwise ungrounded. According to theorists, Walster and colleagues, the four propositions are:

I. Individuals will try to maximize their outcomes (where outcomes equal rewards minus punishments).

II. a. Groups can maximize collective reward by evolving accepted systems for equitably apportioning rewards and punishments among members. Thus, groups will evolve such systems of Equity and will attempt to induce members to accept and adhere to them.

b. Groups will generally reward members who treat others equitably, and generally punish members who treat others inequitably.

III. When individuals find themselves participating in inequitable relationships they will become distressed. The more inequitable the relationship, the more distress they will feel.

IV. Individuals who discover they are in an inequitable relationship will attempt to eliminate their distress by restoring equity. The greater the inequity that exists,

[27]Equity theory has, of course, a long history and extensive bibliography in moral and economic philosophy that cannot be dealt with here. For a recent treatment that attempts to deal with everyday situations see Young (1994).

the more distress they will feel, and the harder they will try to restore equity (Walster, et al. 1978: 6; see also, Donnerstein and Hatfield 1982).

Walster, et al. state unequivocally that equity theory rests upon what they consider "the simple, but eminently safe assumption that man is selfish" (1978: 7). Except for the fact that the theorists remain in the rational choice, self-interested framework, their propositions, as asserted, are expressive of the reciprocal algorithms of behavior.

The apparent logical inconsistency of proposition 1, maximizing outcomes, with proposition 4, restoring equity, is glaring under the assumption that man is selfish and seeks to maximize. If man selfishly seeks to maximize, why should he be satisfied with merely restoring equity rather than seeking to reverse the situation to one of dominance in his own favor? This logical inconsistency of the propositions becomes resolved when the tug of empathy is added to self-interest. According to the reciprocal algorithms of behavior, empathy is what permits us to settle, without excessive distress, for a position of equity rather than a reversal of the inequities of the relationship which would be required under a concept of maximizing.

I might reiterate with confidence, then, that reciprocity, proceeding from the evolved structure of our human brain, is the basic structuring dynamic of our social lives. Anthropological and social research have convincingly shown that reciprocity underpins our most primitive and basic social relations, from family interactions, to gift exchange, to the foundations of more complex economic life. It continues to do so in a less obvious, but no less pervasive manner in our modern systems of exchange.

Chapter 10

Political Economy: The Reciprocal Brain and the Management and Creation of Scarcity

Human society is a product of the human brain. The aspects of society that we term political and economic are likewise products of the human brain. There is no other source for them. There are no political or economic essences or universals, independent of the human brain, existing out there in a positivist, mechanical world waiting to be discovered.

There are, however, environmental constraints. And the brain evolved in dynamic interaction with these constraints. Such constraints include not only the basic constraints of food, shelter, safety, but also the social constraints of the mammalian life form. The human brain functions, among other things, as a normative, evaluative, and environment-shaping organ based upon its evolved mechanisms to assure survival of the individual and the species within the existing constraints. This is accomplished by the dynamic programming for self-preservation and affection, ego and empathy, self- and other-interest.

All human politics and economics are manifestations of the normative, evaluative functioning of the human brain based upon this dynamic programming. There is no such thing as a positivist, value-free human politics, economics, or any other aspect of society.[28] There are only normative variants. Without referring to brain science, I take it that this is what Nobel laureate James M. Buchanan is saying when he writes about the inevitable normative aspects of economics:

> ...And let us be sure to understand that there is no "is" that is "out there" to the observing eye, ear, or skin. We create our understanding of the "is" by imposing an abstract structure on observed events. And it is this understanding that defines for us the effective limits of the feasible. It is dangerous nonsense to think that we do or can do otherwise (1991:41).

These normative variants are the range of possible expressions of the algorithms of reciprocal behavior of our evolved brain structure acting within and upon

[28]The value-free claims of economics and economists are becoming more and more difficult to sustain. The appreciation of this difficulty has led to an increased interest in ethics and economics (e.g., see Wilber 1998:2)

environmental constraints...many of them created by the brain itself, interacting of course with other like brains.

The environmental constraints, then, are the original ones constraining basic survival plus those modifications and additions created by the reciprocal dynamic itself as it shaped our social environment as well as altered our physical environment. Our social, political and economic traditions, institutions, and practices, to include the changes that we have made to our physical resources in the way of technology, are the products of our brain dynamic acting within and upon existing environmental constraints.

One of the greatest shortcomings of economic science to this date has been its failure to appreciate adequately the basic, prime moving, shaping power of the human brain in all things economic and political...as if there could be any other source for such phenomena. Thorstein Veblen, writing earlier, caught the essence of the problem in an eloquent and often quoted passage which described the self-interested or hedonistic conception of man as without antecedent or consequence, buffeted about by forces that push him mindlessly in one direction or another. He concluded that "Spiritually, ***the hedonistic man is not a prime mover***" (emphasis mine) (Veblen 1948: 232-233).

Although Veblen defined the problem of the passive economic man, he got caught up in the materialism and the emergent Pavlovian-Watsonian simplistic stimulus-response behaviorism of his day and merely substituted it for the Newtonian model of physical forces acting on the still passive economic man.

Despite the more recent addition of the concept of a more dynamic, wealth-maximizing rational economic man, influenced even more recently by the addition of such neoclassically ignored factors as habit, convention, and institutions, which bring history to bear on the neoclassical isolated moment of economic decision, the economic man is still treated largely passively. This passivity is achieved by the imposition of narrow constraints that obscure the active, shaping dynamic of the brain and force the so-called rational economic actor into equally narrow pre-set options of behavior. Even in the more process-oriented, evolutionary approach of the new institutional economics, the dynamic shaping power of the brain is not adequately appreciated (e.g., see Langlois 1986). Among the public choice and institutionalist economists, A. Allan Schmid gets close to a grasp of the reciprocal algorithms of behavior in his focus on the market as a set of interrelationships. He writes: "Rights and opportunity sets are seen as ***reciprocal*** (emphasis mine), where one person's freedom to act is another's limitation." (1987: xiv). There are important differences, however. In his psychological interpretation of the highly generalized concept of utility maximization, self-interest and altruism are seen as static or stable preferences rather than as the shaping dynamic proceeding from our evolved brain structure (1987: 197-206). This view of egoism, altruism, as well as all other behavioral attitudes, as nondynamic, stable preferences characterizes economic thinking and allows the broad economic approach to be seen as providing a comprehensive framework for understanding all human behavior (e.g., Becker 1987; Frey 1992).

The laws, principles, regularities, and the dynamics of politics, economics, and social life itself simply do not exist and cannot be understood independently of the human brain. Any presumed detachment from them is illusory. This holds even in the most rarefied applications of cliometrics and econometrics. Pull the human actors out and the dynamics and econometrics will simply disappear.

When researchers claim they are using positivist, objective approaches, they are simply deluding themselves. A positivist approach in politics or economics usually means that the researcher is implicitly accepting the pre-existing and undefined normative structure of whatever he/she is researching and then assuming that structure away. It is a fundamentally fallacious posture, although it may be assumed, with highly constrained plausibility, in the short term as an operational research position of convenience in order to reduce complexity. Nevertheless all findings or interpretation of findings above the level of triviality will inevitably fall back upon the denied normative foundations or an alternatively substituted set.

NATURAL ENVIRONMENTAL CONSTRAINTS AND THE PROBLEM OF SCARCITY

The fundamental natural environmental constraint is scarcity. Life exists everywhere in a state of scarcity. The state of scarcity does not mean that the environment is hostile to life. And that life must struggle for survival against this fundamental hostility. The natural environment is, on the contrary, very supportive of life. Early fragile life forms would never have thrived and evolved the amazing variety and complexity of today if the natural physical environment were not extremely supportive.

Although the environment does have limits, this scarcity is created by the nature of the organism itself. In this rich, supportive physical environment, the organism, left to its own processes, tends relentlessly to increase to the carrying capacity of the environment. This is a fundamental biological principle. Life is resilient, relentless, blindly automatic, proliferating endlessly within the environment until it runs up against the limits of the environment's capacity to support it. Life inevitably, by its very nature, creates the constraint of scarcity.

We humans, like all other forms of life, create our own constraint of scarcity. This inevitable creation of scarcity by life is the most fundamental, shaping factor of what we call economics. The survival strategy of the human species, like any other organism, is the set of characteristics and behaviors the species evolves or develops for coping with the constraint of scarcity that it produces. This fundamental biological principle, followed also by the human species, that a species will expand to the carrying capacity of the environment, sets the Malthusian tone of economics as the dismal science.[29]

[29]Scarcity is seen by economists and presented in introductory texts as the basic economic problem (e.g., see Kohler 1992: 3; 1968: 5; Allsopp 1995: 11-29). Economist and Nobel laureate Gary Becker sees the definition of economics in its broadest and most general terms as concerned with scarce means and competing ends (1987). This reflects the influential definition given by Robbins first in 1932. Robbins defined economics as "the science which studies human behavior as a relationship between ends and scarce means which have alternative uses" (1952: 16). Arndt, however, calls scarcity the Cinderella of economic theory because while recognized as basic to economics, it is excluded from effective consideration in economic theory through a series of assumptions including that of equilibria and the more dubious assumption that all scarcity can be overcome by exertion of human industry...in other

The human brain evolved as a scarcity coping organ in a primarily kinship based foraging society where sharing or reciprocity was essential to survival and which reinforced the adaptive evolution of the mammalian characteristics of self-preservation and affection (e.g., see Isaac 1978; Tooby and Devore 1987; Cosmides and Tooby 1989: 59; Cummins 1998). The algorithms of reciprocal behavior of our brain structure are, likewise, a scarcity-coping dynamic. They are the dynamic by which human society manages, to the extent that it does, the constraint of its self-produced scarcity.

TECHNOLOGY--COPING WITH NATURAL CONSTRAINTS

The tools, skills, and knowledge which humans create for coping with the natural constraints of the physical environment, as well as later devised social constraints, constitute technology. The evolution of the more recent neocortex gives us the capacity, unlike other species, to alter many of the constraints imposed by nature. Most significantly, we can and have altered the carrying capacity of the environment. For example, we have increased food production and distribution and we have increased living space by erecting buildings skyward. The fundamental thrust of the technology of production is to increase the carrying capacity of the environment, to overcome the constraint of scarcity that we ourselves produce in the first place. Productivity itself has a circular, self-reinforcing effect. The logic of productivity carried to its logical extreme as a limitless end in itself plays upon the scarcity-coping dynamic of our brain structure and turns it, further, into a scarcity-generating dynamic. I will return to this thought later on.

Technology allows us to alter the carrying capacity of the environment. It does not, however, eliminate the inevitable self-generated constraint of scarcity. Instead, it sets up an arms race with an uncertain outcome. As our technology increases the carrying capacity of the environment, we have the inevitable biological tendency to increase in numbers inexorably to reach the limit of that capacity. Ultimately, we will either control by the rational power of our evolved brain this previously inevitable biological principle, or we will continue to increase the carrying capacity of the environment infinitely, or we will exhaust the resources available to us even for additional technology. In the end we will either overcome, control, or become victim to our self-generating constraint of scarcity.

DERIVED ENVIRONMENTAL CONSTRAINTS--INSTITUTIONS

Within this self-produced scarcity-constrained natural environment, humans devise additional constraints. Such devised constraints are called institutions in the economic literature (e.g., North 1990). These institutions are constraining rules or principles for dealing ultimately with the natural and fundamental constraint of scarcity. Since our three-level brain mechanism is our fundamental evolved social mechanism for dealing with scarcity, these institutions, rules, or devised constraints invariably regulate or order our evolved dynamic of reciprocity.

---

words, that there are no limits to productivity (1984: 17-36). On the latter theme according to Simon (1981) modern resource economics treats limits as a short-term constraint that a dynamic economy will overcome. In this view neither resources nor any other form of limit is seen to ultimately constrain what the competitive economy can do. Schmid sees scarcity as a function of nature and human tastes (1987: 8).

BETWEEN ECONOMICS AND POLITICS

In recent years the new field of positive political economy has emerged from advances in interdisciplinary research in economics and politics. It is explicitly theoretical seeking to discover principles and propositions by which to compare, explain, and understand actual economic and political experience. It is micro-economic in its focus and is grounded in the rational actor methodology, applying the assumption of constrained maximizing and strategic behavior by self-interested agents to explain the origins and persistence of political institutions as well as public policy (Alt and Shepsle 1990: 1). Its claim to be "positive", however, is suspect because it rests on the acknowledged evaluative, normative assumption of self-interested, wealth-maximizing individuals. In its failure to recognize the reciprocal nature of exchange as primary, its findings will always be distorted by its underlying assumption of exclusive self-interest. Reciprocity, in the absence of its recognition as the proper basis for microeconomics, will nevertheless be dealt with pervasively and implicitly. Only the fundamental, shaping dynamic of movement and change will be obscured...as it has and continues to be in microeconomics.

The irony and contradiction of the emphasis on self-interested, wealth maximizing behavior by the new positive political economy, which denies empathy or benevolence its place in the reciprocal dynamic, is clearly reflected in the survey introduction to the new subject by Alt and Shepsle (1990). Whereas in the introduction the editors unequivocally claim, as pointed out above, that the theory and methodology is based on the wealth-maximizing self-interested rational actor, in a later included article Nobel Prize winner Gordon Tullock begins by telling a story about the first time he saw another famed Nobel prize winner, Milton Friedman, at a public debate of free enterprise versus socialism. According to Tullock, in that debate Friedman based his entire lecture on what a ***benevolent*** (emphasis mine) dictator would do. Tullock explains that Friedman intended this simply as a rhetorical device to argue for a free economy and was doing what all economists of that time did..."investigated optimal policies and considered what ***well-intentioned*** (emphasis mine) people would do if they had control of the government" (Tullock 1990: 195). At the conclusion of the same article Tullock again acknowledges inadvertently the pervasive factor of empathy when he writes: "Almost all economists, whatever they say, are actually reformers who would like to improve the world."(1990: 210).

In other words, as indicated plainly by Tullock, the finest and foremost in the field of economics are always motivated by and act on empathy, benevolence, and good-intentions rather than the wealth-maximizing self-interested motive ascribed to all actors by their theory and methods. These economic theorists are either seeing themselves with implicit arrogance as possessing morality superior to the motives of the lessor rational economic man that they deign to study. Or else, caught up in the rhetoric of self-interest, they implicitly acknowledge, but are overtly and explicitly blind, to the pervasive function of empathy in the socio-economic-politico process of exchange.

COST/BENEFIT ANALYSIS AND HUMAN RECIPROCITY

At this point it is useful to reemphasize that every human relationship or interpersonal act is a social exchange relationship or act. Every such relationship or act can be analyzed in economic terms on a cost/benefit basis or rather in cost/benefit terms. Cost/benefit is the economic vocabulary for the give and take, the empathy and ego of human reciprocity. Cost/benefit carries the positivist illusion of objectivity because of the association with mathematics, things that can be counted, quantified.

Cost and benefit, what we give and what we take (or get), are just alternative ways of talking about reciprocity, the tug and pull of ego and empathy. We demand and take for our ego, we provide (supply) and give through the vehicle of our empathy. Without empathy we could not engage in a social exchange, economic or otherwise. We would lack the basis for recognizing, responding, and supplying to the needs or demands of others.

Viewing reciprocity in the artificially detached, quantified cost/benefit terms of current economics may be useful in the counting house. And there need be no overly compelling objection to such as view, held temporarily as an operational convenience, as long as it has some utility and it doesn't claim or substitute itself, explicitly or implicitly, as a full and complete representation of human reciprocity. Unfortunately, such a view inevitably gravitates to that latter position...which is expressed best in the simplistic, reductionist, dehumanizing assumption that pervades economics...the assumption of a solely wealth-maximizing self-interested individual economic man. This pervasive assumption obscures or denies the true nature of human reciprocity and has the effect of distorting and dehumanizing social, economic, and political exchange at all levels of the process.

POLITICAL ECONOMY AND THE MANAGEMENT OF BEHAVIORAL TENSION

The new emphasis on the unity of economics and politics, however, is appropriate and welcome. Broadly speaking it may be considered that institutions concerned with order are political and those concerned with reciprocity and exchange are economic. There is, however, no clear separation between the two. Where they merge into each other, we have institutions of political economy.

To the extent that institutions provide order (regulate), and they invariably do, they are inherently political. To the extent that they are social, and they invariably are, they impact reciprocity and are therefore economic. Most institutions, whether in the form of principles or whether expressed in concrete organizations reflect this duality. They are concerned with ordering reciprocity in some way. And in that sense they are politico-economic.

The institution of private property, or individual property rights, for instance, has this dual function. It is political in that it provides for the maintenance of a personal and protected resource base for the economic purposes of survival and reciprocal exchange. According to the algorithms of reciprocal behavior, to the extent that unequal holdings of private property or property rights are permitted in any society, the institution of private property produces unbalanced reciprocity among members of the society. Unbalanced reciprocity, potential or actual, leads invariably to inequalities of power. Schmid, for instance, bases virtually his entire concept of power on the presence or absence of property rights.

> One's right is another's cost. One person's property right is the ability to coerce another by withholding what the other wants...To own is to have the right to coerce (1987: 9).

And further on he writes:

> Power is inevitable if interests conflict. If everyone cannot have what they want simultaneously, the choice is not power or no power, but who has the power (1987: 9).[30]

Power is another way of speaking of hierarchy or relationships of dominance or submission. The algorithms of reciprocal behavior dictate that such relationships always and invariably carry a degree of behavioral tension to the extent of the imbalance.

Unbalanced reciprocity, hierarchy, power, inequality, dominance and submission are all aspects of the same phenomenon, the dynamic of the reciprocal algorithms of our evolved brain structure. And wherever and whenever any of these aspects are manifest, there will be an accompanying proportion of behavioral tension.

This is not to argue against private property. Private property rights are the basis of the free enterprise economy which is, at this point in our economic evolution, essential to our democratic political processes. The purpose here is to acknowledge the behavioral tension of inequalities in the holding of property rights. Such behavioral tension, such inequalities, constitute one of the major management problems of political economy.

[30]See Barlett (1989) for the development and application of an economic theory of power which accepts scarcity as a given and maximizing utility as the driving assumption.

Chapter 11

Institutions, Organizations, and Reciprocity

In primitive societies, or in the more natural state of humankind, unbalanced reciprocity may result from natural differentials in strength, talent, ability, or intelligence. In simple societies the hierarchies created by such natural differentials will be inherently unstable because of behavioral tension and will shift as the differentials become less so.

In more complex societies, economic or political institutions, whether expressed as principles or concrete organizations, order or regulate reciprocity to some degree. In doing so they enter into the behavioral conflict of the tug and pull of ego and empathy, self- and other- interest, the give and take among the members of society. All such institutions and organizations, then, carry an inherent load of behavioral tension. They also add additional costs to the exchange process. Such costs are referred to in economics as transaction costs.[31] Such transaction costs are an index of the behavioral tension added by the institutions.

Consider for instance the institution of private property. As noted, that institution attempts to order (regulate and systematize) reciprocity by permitting and providing for the maintenance of a personal and protected resource base for survival and for reciprocal exchange. Such an institution may serve to mitigate natural differentials.

[31] The focus on transaction costs in the economic theory of institutions derives from two articles by Ronald H. Coase on the nature of the firm or business organization and on social cost (1937, 1960). The interest in transaction costs was also abetted by the work of George J. Stigler (1961) and Friedrich A. Hayek(1937, 1945). The work of Oliver E. Williamson (1975, 1985, 1991, 1996) led to the full development of transaction cost economics treated more fully in chapter 12 of the present work. Institutions are seen by Williamson as created to ***reduce*** transaction costs. But such a view assumes the preexistence of a market. The market and associated transaction costs develop in a mutually reinforcing feedback relationship. Once the market is established, the question becomes which among the institutional alternatives carries the lower transaction costs. Or can we create institutions that carry lower transaction costs? My point here is that all institutions inherently carry transaction costs. Those that carry the lesser costs can be said to reduce costs within an existing state of the market.

From the standpoint of the survivability of the society, however, we must consider more long-term effects. As noted earlier, the logic of productivity carried to its logical extreme as a limitless end in itself draws upon and reinforces the scarcity-coping dynamic of our triune brain structure and turns it into a scarcity-generating dynamic. The capitalist system as an institution or set of institutions is designed to do just exactly this. The generation of scarcity is captured in the ever popular phrase, *the creation of demand.* The major thrust of the advertising industry, which supports that logic, is unabashedly the creation of demand. The created demand is, of course, to be responded to by a reciprocal of newly created supply. The engine of reciprocity based upon our reciprocal algorithms of behavior, primed and reinforced, grinds away relentlessly...creating demand, generating scarcity. Escalating increase of productivity becomes an end in itself.[34] This, in fact, has been the thrust of western and, particularly, Anglo-American economic theory. It underpins all economically-based theories of growth and development. And it is underpinned by the normative assumption that ***all*** productivity is good. The more of it the better. Standard economic theory assumes no negatives to what is produced and no limits to the growth of productivity. It is an amazingly naive and simplistic normative paradigm.

DESIRABLE VS SUSTAINABLE LEVELS OF PRODUCTIVITY

A dilemma, intruding to some degree these days in all societal level thinking about productivity, and looming ominously on the horizon, is the question of desirable versus sustainable levels of productivity. A major question facing us ultimately (if and when we have achieved the maximum desirable or sustainable productive state or equilibrium) is: what will be the fate of capitalism or the nature of the economy at that time? The reciprocal dynamic of our evolved brain structure will still be with us. We may need to find other ways to express our social reciprocity than the endless and mindless production of goods and services. Will we find other effective and satisfying ways of expressing our reciprocal sociality in more balanced and humane forms of community? Will we shift more to reciprocity in esthetics, morality, intellectual expression, or even spirituality?

But we are not there yet, from either the standpoint of desirability or sustainability. Most clearly, we have certainly not reached the maximum desirable level of productivity. From a world viewpoint, we still have considerable deficits in survival essentials (e.g., food, shelter, and health care) and even greater differentials in their balanced reciprocity throughout the world. This is the major source of behavioral tension within and among nations. From the standpoint of sustainability, the other side of the question we must confront, the answer is not so clearly cut. There are, however, warning signs aplenty in the environment. When confronted with the threat of a limit to maximum sustainability, under the existing assumptions of our economic theory, we will depend upon the further evolution and development of our technology to move the threat further toward the distant horizon...to increase and extend the carrying capacity of the environment as we continue to generate the constraints of scarcity. The concern over the question of sustainability has produced

[34]Compare the discussion in Power (1996: 211-13). Without reference to the dynamics of brain structure, Power captures the social effects descriptively in his section on the treadmill of competitive consumption.

a new subdiscipline of economics, called ecological (or sometimes, living) economics (e.g., see Daly 1980; Daly and Cobb 1989; Ekins 1986; Ekins and Max-Neef 1992).

It is urgent in any case to understand the dynamic of our brain structure. And develop the ability and the sense of urgency and responsibility to manage it. The logic of the present ***relentless and endless production as an end in itself*** economic system, spreading now to all corners of the globe, if it cannot be managed wisely, or if it spins out of control, may lead us to the brink of extinction as a species by exhausting the carrying capacity of the environment. We are not there yet, but it is time to take very seriously the responsibility for management and to consider modifications and alternatives.

Our brain structure has endowed us with a two-horned dilemma. When we crank up the productivity engine, at the same time we inevitably crank up the scarcity-generating engine. When we full throttle the engine under the governing logic of relentless and endless production as an end in itself, we also full throttle the relentless and endless generation of scarcity.

There are three possible outcomes:

1. We will exhaust the ultimate carrying capacity of the environment and go extinct.

2. We will reach a compelled, but managed, equilibrium and survive by recognizing and accommodating the limits of sustainability.

3. We will continue to develop new technologies that will infinitely extend the carrying capacity of the environment, perhaps extending it into the reaches of space, which we will then colonize.

The final answer concerning each of these possible outcome scenarios is uncertain and, at our current state of knowledge, is impossible to predict. The one out of the three that we as individuals choose to consider to be most likely depends on whether we are disposed to be pessimistic or optimistic. The pessimistic position on the issue of sustainability is eloquently covered in the literature on population and ecology (e.g., Ehrlich 1969; Ehrlich and Ehrlich 1990, 1996; Mazur 1994; and somewhat more hopefully, Chertow and Esty 1996). Crispin Tickell, longtime advisor to successive British prime ministers and former president of the Royal Geographical Society presents an elegant overview of the issues in an article titled provocatively, "The Human Species: A Suicidal Success?" which appeared first in *Geographical Journal* (1993) and is repeated as the concluding chapter in *The Human Impact Reader* (Goudie, 1997). The optimistic position is mindlessly fueled,[35] implicitly if not explicitly, by the several varieties of mainstream economics which assume away the basic economic problem of scarcity or limitations (e.g., see the criticism in Arndt 1984: 17-36) and seek endlessly increasing productivity based upon unbalanced appeal to the wealth-maximizing self-interested side of human nature ... the so-called "rational" economic man.

[35] Economist Julian Simon argues the most straightforward and unabashed case for our ability to overcome any scarcity problem by full throttling our productive economy (Simon 1981). See also his debate with Norman Myers, who argues the opposite case (Myers and Simon 1994, especially, p. xv).

Wealth-maximizing self-interest may be anything but rational in the scenarios that are developing before us in the not-to-distant future.

In the interim, however, while we grope for a better balance between self-interested, egoistic consumerism and other-interested, empathetic social responsibility, we must manage politically the inevitable behavioral tension that exists and develops within and among nations, so that we do not destroy ourselves prematurely, along the way to one outcome or the other. This is the short- to mid-range challenge of both economics and politics.

Chapter 12

The New Institutional Economics: Williamson and Transaction Cost Economics

The emerging study of new institutional economics aims at the integration of the neoclassical economic theory with institutional theory.

Neoclassical economic theory broke from the somewhat ad hoc economic history of the eighteenth and nineteenth centuries and applied to that history a systematic body of theory buttressed by sophisticated quantitative methods. The motivation for neoclassical theory was the search for principles and regularities that could apply generally to economic analysis independent of history. Of course, once you find universal or general principles and regularities that operate independently of history, history becomes unnecessary or irrelevant...except perhaps, as an absorbingly interesting human social record. Neoclassical theory, with its history-independent principles and regularities is further ahistorical in that it focuses on the allocation of resources at a moment in time, rather than such allocation over more extended periods of time. In that sense, then, ahistorical neoclassical economic theory effectively killed its historical subject matter. The methodical application of price theory to economic history was, however, undoubtedly a major contribution producing important insights. Nevertheless, it was the application of an ahistorical theory and methodology to a historical subject matter and it therefore obscured as much as it elucidated.

But neoclassical theory obscured even more than history. In its focus on principles and regularities the effects of institutions were also assumed away. This allowed the principles and regularities to operate in what is called a frictionless or socially cost-free environment. In effect, then, the theory assumed away both history and social context.

This assuming away of both history and social context left neoclassical theory with no effective way to explain change, differential performance, or the decay of economic societies over time.

But there was even a third limiting factor...one that has never been effectively resolved. That factor is the assumption of an exclusively self-interested, wealth-maximizing human nature. This assumption made accounting for the pervasive factor of human cooperation, so fundamental to the maintenance and function of any society, difficult and awkward to explain. The new institutional economics attempts to deal with the problem of cooperation. Leading thinkers Oliver Williamson and Douglass North approach the problem from different perspectives. This chapter deals with the new institutional economics from Williamson's perspective. The next chapter looks at the same subject from the perspective of Douglass North.

WILLIAMSON'S NEW INSTITUTIONAL ECONOMICS

The new institutional economics, popularized as a term by Williamson (1975), harks back to the former ad hoc, more broadly based institutional analysis and aims at overcoming the shortfalls of neoclassical analysis by bringing back in the historical perspective plus the constraining and shaping effect of institutions. Nevertheless, as Langlois (1986: 2-5) points out the new version, although sharing the concern for institutions with the earlier American institutionalists, is not historically or conceptually continuous with that tradition, but rather may owe more to their opponents, especially the Austrian economist, Carl Menger. Williamson's study focused on markets and hierarchies. Hierarchies refer to institutions and organizations. Williamson, in this work, never makes a clear distinction between the two.[36]

In his pioneering study of the new institutional approach to markets and hierarchies, Williamson maintains the economic assumption of wealth-maximizing self-interested individuals. To this he adds the further emphasis of opportunism, which means "self-interest practiced with guile"(1975: 26) or deceit. He sees the control function of hierarchies aimed partly at restraining this more blatant aspect of self-interest, opportunism.

One must look very closely to grasp the implicit, pervasive ground of Williamson's analysis because he spends the greater part of his time on exceptions, or barriers, to effective cooperation that he wishes to control or overcome. This implicit ground is, nevertheless, identifiable as the reciprocity of ego and empathy. This is revealed in the following statement on what Williamson considers "attitudinal considerations."

> The problem in all this is to identify when such attitudinal considerations operate strongly and when they can be safely neglected. I conjecture that transactions which affect conceptions of ***self-esteem*** (emphasis mine, read self-esteem as ego) and/or conceptions of ***collective well-being*** (emphasis mine, read collective well-being as a function of empathy) are those for which attitudinal considerations are especially important... Individuals who experience nonmetered externalities in noneconomic

[36]Bromley sees the failure to distinguish between institutions as rules of organizations and as the organizations themselves as the source of considerable confusion in the new institutional literature. The term institution is variously used to refer to organizations, e.g., banks as financial institutions; to a person or position, e.g., the presidency; and to the rules defining economic relations between individuals, e.g., private property (1989: 27-8). A somewhat cleaner distinction is offered by Davis and North who define the institutional environment as the set of political, social, and legal ground rules that govern economic and political activity (1970: 133). Schotter (1981, p. 11) defines an institution as follows: "A social institution is a regularity in social behavior that is agreed to by all members of society, specifies behavior in specific, recurrent situations, and is either self-policed or policed by some external authority." See also North 1990.

contexts and reach nonmarket accommodations thereto bring the attitudes and experiences which evolve in these nonmarket circumstances to the workplace as well. Rather than regard transactions in strictly ***quid pro quo*** (emphasis in original) terms, with each account to be settled separately, they look instead for a favorable balance among a related set of transactions (1975: 256-57).

The above qualifying statement in the concluding pages of Williamson's analysis reveals clearly the assumed reciprocal dynamic of ego and empathy, self and other interest, that plays implicitly at the foundation of Williamson's analysis of organizational market hierarchies.

The underlying reciprocal ground of Williamson's study comes out most clearly in his chapters on peer groups, simple hierarchy, and the employment relation (1975: 41-81). In assessing the worth of worker peer groups, he writes:

> Associational benefits can accrue to peer groups through increased productivity among members of the group who feel a ***sense of responsibility*** (emphasis mine) to do their ***fair share*** (emphasis mine) as members of a group but, left to their own devices, would slack off. Also, and more important, a ***transformation of "involvement" relations*** (quotes in original, emphasis mine), from a ***calculative*** (emphasis mine) to a more nearly ***quasimoral*** (emphasis mine) mode, obtains (1975: 44).[37]

The use of the terms ***sense of responsibility*** and ***fair share*** are judgmental statements incompatible with the assumption of an exclusively self-interested, wealth-maximizing individual, but fully compatible with an individual experiencing the tug and pull of ego and empathy, the reciprocal algorithms of behavior. Followed in the next sentence by the transforming effect of ***involvement relations***, which term clearly refers to empathetic social relations, and the movement from a self-interested ***calculative*** to a more empathetic ***quasimoral*** mode, these statements confirm the implicit and unclarified assumption of reciprocity in contradiction to the stated one of self-interested wealth-maximizing economic man.

Further on, after discussing why hierarchical organization can overcome the shortfalls of peer groups by controlling opportunism and accomplishing more efficient decision-making, Williamson writes:

> What then prevents the peer group from being displaced by simple hierarchy? The main reason, I submit, is that peer groups afford ***valued involvement relations that are upset*** (emphasis mine), in some degree, by hierarchy. Not only is transparent ***inequality of rank*** (emphasis mine) considered objectionable by some individuals, but auditing and experience-rating may offend their sense of ***individual and collective well-being*** (emphasis mine: read ego and empathy, self- and other-interest) (1975: 55).

This passage, like the previous one, not only confirms the implicit dynamic of reciprocity, but it also implicitly acknowledges the role of behavioral tension accompanying the tug and pull between ego and empathy, self- and other-interest. It does so by describing involvement relations as being upset by hierarchy and inequality of rank as objectionable and offending the sense of individual as well as collective well-being. Why on earth should exclusively self-interested, wealth-maximizing humans be upset by the effects of hierarchy and inequality on collective well-being?

[37]In his later work, Williamson observes that when managers stop playing ... that is, behave ethically (empathically, morally)...it reduces transaction costs. For example, he refers to the Japanese case, writing that transaction costs are reduced in Japan because Japan has institutional and cultural checks on opportunism (1985: 122).

In discussing the efficiency challenges of collective organization, Williamson tells us at one point that:

> ...Self-enforcement is tantamount to denying that human agents are prone to be opportunists, and ***fails for want of reality testing*** (emphasis mine)... 1975: 69)

and at another point that:

> ...To observe that the pursuit of perceived individual interests can sometimes lead to defective collective outcomes is scarcely novel...An enforceable social contract which imposes a ***cooperative*** (emphasis mine) solution on the system is needed...both private collective action (of which the firm, with its hierarchical controls, is an example) (parentheses in original) and ***norms of cooperation*** (emphasis mine) are also devices for realizing ***cooperative solutions*** (emphasis mine)... (1975:73).

And in a following paragraph, he adds:

> Although it is in the interest of each worker, bargaining individually or as a part of a small team, to acquire and exploit monopoly positions, it is plainly not in the interest of the *system* (italics in original) that employees should behave in this way...what this suggests is that the ***employment relations be transformed*** (emphasis mine) in such a way that systems concerns are made more fully to prevail and that the following objectives are realized...***consummate*** (emphasis mine) rather than ***perfunctory cooperation*** (emphasis mine) is encouraged...(1975:73).

When Williamson claims that self-enforcement fails for want of reality testing, going on to say that ***cooperation*** must be ***imposed*** by an ***enforceable social contract*** or perhaps by ***norms of cooperation***, he is vacillating between assumptions of self-interest and reciprocity. When he refers to the conflict between system and worker interests, he is fully within the self-interested, wealth-maximizing framework. When he aims at ***transforming employment relations*** so that workers ***consummately*** (wholeheartedly and empathetically, because it's the right moral thing to do) suppress their own self-interested maximizing rather than ***perfunctorily*** (minimally to keep from getting fired) so cooperating, he is either engaging in wishful speculation, attempting to transform human nature as understood by economics, or grasping implicitly and hopefully at the intuited but unarticulated dynamic of reciprocity.

Williamson's analysis may be viewed, contrary to his explicit stance, as largely an analysis of exceptions. It is focused upon exceptions and the control of deviations that disrupt this implicitly hoped for reciprocity and add transaction costs to the process of exchange...to the desired and implicitly expected identity and commitment to the collective welfare. His definition of opportunism (which hierarchy is designed to control) as self-interest with guile or deceit is a loaded term which carries the unspoken, unclarified fundamental assumption that self-interest must have its limits. In other words, the definition contradicts the individual wealth-maximizing self-interested assumption of microeconomics. When self-interest reaches the border of those tolerable limits, empathy must set in to balance in the form of commitment to collective welfare.

The emphasis on self-interest and its deceitful extreme, opportunism, has become a fundamental of Williamson's theory as it has continued to develop along the lines of the earlier work (e.g., 1985; 1991; 1996). The focus on the transaction as the basic unit of analysis and the increased transaction costs associated with the assumption of ever-present opportunistic behavior has stamped the mainstream of the new institutional economics, which is now more frequently referred to as transaction cost economics (TCE) (e.g. Groenewegen 1997). The truth is that

Williamson (as well as other transactional researchers), despite his negative emphasis on the control of opportunism, operates on the implicit premise that the basic purpose of business organization and organizational behavior is to achieve the collective welfare of the firm by rewarding the self-interest of individual members and, in return, capturing the empathetic behavioral reciprocal of the same individuals in the form of identity with the firm and commitment to the firm's market objectives and collective welfare. In another context not directly related to TCE, Hampden-Turner and Trompenaars make essentially this same point when they write

> Another vital requirement of all work organizations is the provision of care, attention, information, and support to each of its individual members ... It is an underlying condition of the success of an enterprise that the individual's initiative, drive, and energy be harnessed to the purposes of the organization (1993: 8).

The reciprocal algorithms of behavior, then, are clearly present and fundamental. They are obscured by positivistic assumptions and the jargon of a pretentiously and speciously "objective" approach to microeconomics. But more important than that, the analysis is obscured by the contradictory nature of the paradigm. Under a socio-economic system that has as its fundamental goal the profit-maximizing collective organization or firm, backed by the equally fundamental assumption of the wealth-maximizing, exclusively self-interested individual, it becomes glaringly obvious, in fact, that one set of framework and rules is being applied to the system and another contrary set is pressed upon the individual worker as a part of the system.

There are in fact two different paradigms being applied, one to system or firm, another to worker or individual. The faulty and contradictory paradigms, as applied, obscure and frustrate the necessary reciprocity and generate high levels of behavioral tension that impede effective management. Thus we have hi-lited some of the frustrating and embarrassing Ptolemaic epicyclic-like exceptions that spin off from the assumption of the exclusively self-interested, wealth-maximizing economic man that underpins and pervades the paradigm of our current economic science. Simply acknowledging the reciprocal algorithms of human exchange behavior would go a long way toward eliminating the exceptions as well as understanding and properly accounting for the dynamics of economic exchange.

THE TRANSACTION AS THE BASIC UNIT OF ANALYSIS

There is a further problem with current transaction cost economics (TCE). TCE fails to grasp the true nature of its self-proclaimed basic unit of analysis, the transaction. According to Williamson, emphasis on the basic transactional unit is seen to be a distinguishing feature of TCE vis-à-vis more traditional approaches (1996: 6). Part of the confusion in current TCE thinking is to be found in the failure to properly grasp the nature of the basic atom of transaction. This failure is likely brought on by having proceeded from within the long-standing, exclusively self-interested behavioral assumption of classical economic theory. This faulty assumption has long obscured and distorted the true nature of the market, causing self-interest rather than reciprocity and cooperation to be seen as central to market function. It is a complete misplacement or misstatement of the reality of the market.

The ***basic unit of analysis***, the ***transaction***, is itself fundamentally a ***unit of reciprocity***, of cooperation, of the tug and pull of ego and empathy. The transactional atom when opened up or unpacked consists of the two elements, ego and empathy in a state of negotiated tension or cooperation. Reciprocity, or cooperation, then is the over-arching, all-bracing essence of the transaction. Opportunism, or the unbalancing tug toward self-interest, then, is ***deviancy*** within the ***centrality of cooperation and reciprocity***. As illustrated in chapter 8 the transaction evolved from the gift. The transaction, like the gift, is an expression of our mammalian legacy, an act of providing reciprocated by a return act of affirmation with the reciprocal specified to head off the residual tension that produces the added unwanted effect of excessive residual obligation or bonding. TCE's failure to properly define the transaction is in lock step with the failure of received economic theory to properly define the true nature of the market. The exaggerated and inaccurate emphasis on self-interest, opportunism, and greed as the driving forces of the market, rather than cooperation and reciprocity, has served to reinforce and encourage deviancy and give the very valuable institution of free enterprise exchange an undeservedly very bad press.

THE COSTLY PARADOX OF TRANSACTION COST ECONOMICS

One is further led to wonder at the logic of the transaction cost version of the new institutional economics. Since the normative concern of the discipline is, in the name of efficiency, to control and reduce transaction costs caused by self-interested opportunism, why would one want to emphasize self-interested opportunism as the central characteristic of what Williamson comes to call contractual man? The constant self-interested, opportunistic rhetoric fed into the business world through the standard economic literature and in the education and training of economists and managers serves, by well-established principles of behavioral psychology, to perpetuate and reinforce such behavior as central and fundamental.[38] By emphasizing such behavior as central, it becomes the expected, the natural, and, in effect, the encouraged behavior. Cooperative, empathetic, trustworthy behavior then becomes by definition the difficult to be achieved, exceptional behavior that goes against the grain of human nature. This is amazing since cooperative behavior is absolutely pervasive in society and no social organization would be possible without it. The fact that opportunism need not be central to the firm is further evidenced by the Japanese alternative (which Williamson acknowledges), and also the German and French (see Hampden- Turner and Trompenaars 1993). One could as easily and as accurately emphasize empathy, ethics, and moral commitment. Controls could easily be seen as encouraging and reinforcing our natural, pervasive empathetic, cooperative behavior, rather than as limiting pervasive opportunistic behavior.

The reality is that the centrality of self-interested opportunism is not the necessary reality of human nature or Williamson's so-called contractual man, but the self-confirming artifact of the particular paradigm of received economic theory made more extreme by the current, and primarily, American version of transaction cost economics. It is, in fact, a self-defeating, counterproductive artifact because it unnecessarily legitimizes, reinforces, and perpetuates the very transaction-costly

[38]Frank, Gilovich, and Regan(1993) in their article, "Does Studying Economics Inhibit Cooperation," report empirical evidence that exposure to the self-interested model of economics encourages self-interested behavior. See also Maxwell and Ames (1981) and Etzioni(1988).

behavior that it wishes to control and reduce. If current transaction cost economics should shift to a more accurate central concept of reciprocity, the tug and pull of self- and other-interest, it would treat opportunism, not as central, but as deviancy or extreme behavior (which it really is). It would propagate and encourage empathy (trust) to counter that deviancy. By doing so, it would thereby contribute to a normative theoretical position that would reduce the transaction costs rather than adding to them. Niels Noorderhaven (1996), an economist at the Catholic University, Brabant, The Netherlands, has taken a strong step in this direction by arguing for what he calls a split-core model of human nature that includes opportunism and trustworthiness. Although it lacks the flexibility, dynamic and the derivation from neuroscience of the conflict systems neurobehavioral model presented as the foundation of this book and gives too much emphasis to self-interest with deceit (opportunism), the split-core model gets much closer to the underlying reciprocal algorithms of behavior.

Chapter 13

The New Institutional Economics: The Perspective of Douglass North

In the most complete statement of his position, new institutional economist, Douglass North, who sees the problem or question of cooperation more clearly as central, tells us that we must look to two aspects of human behavior to get at the deficiencies of rational choice (wealth maximizing self-interested) theory as it relates to institutional economics. These aspects are motivation and deciphering the environment (1990: 20).

In dealing with the motivational issue, North, in contrast to Williamson, tries to integrate altruism into the calculus. In doing so he draws upon and stays within the externalized, gene-centered work of such sociobiologists as Richard Dawkins of selfish gene fame. His attempted integration of altruism is somewhat forced and unsatisfactory since his externalized perspective does not allow him to enter into the dynamic of motivation. In effect, beyond acknowledging the importance of such behaviors as altruism, he doesn't give us much more in the way of enlightenment on the subject.

In moving on to the question of deciphering the environment, which is the cognitive (as opposed to motivational) issue, North acknowledges the limited capacity of individuals to process adequately all environmental signals and data. In what appears to be an interim or stopgap effort to explain variance in motivation from the perspective of wealth-maximizing behavior, he makes the following statement

> The complexity of the environment, given the limited processing ability of the actor, can explain the ***subjective perceptions*** (emphasis mine) of reality that characterize human understanding and even the ***sense of fairness or unfairness*** (emphasis mine) that the individual feels about the institutional environment... (North 1990: 25).

How "the complexity of the environment" could fully and ultimately explain subjective perceptions and even the sense of fairness or unfairness is very difficult to see. At best such externalities can only partially explain such subjective states. To get a complete explanation one must assume or identify some other motivating sources internal to the human than wealth- maximizing acted upon by, or responding to, the said complexities of the environmental factors. For a sense of justice or fairness one must have in addition to self-interest the capability to identify or empathize with the situation of others...or else the statement is meaningless. To see that just one's self-interested, maximizing self is shortchanged in an economic transaction or situation scarcely constitutes what we consider a sense of fairness or unfairness. One must add to that a capacity for empathizing with others who are shortchanged and perceive that shortchanging of either self or others in the economic transaction or situation is unfair...an instance of unbalanced reciprocity... that offends our human nature.

North acknowledges that his explanation is inadequate to account for a broad range of human behavior (e.g., anonymous free donation of blood, dedication to ideological causes, commitment to religious precepts, self-sacrifice for abstract causes) and that our understanding of motivation is therefore very incomplete (1990:26). After giving this acknowledgment to motivation, he, then, dodges the essential question of motivation and moves on, for further insight, to focus on the study of institutions.

Undoubtedly the examination of institutions needs to be done. And equally undoubtedly, such study will further elucidate the issues of motivation. Nevertheless the project will eventually and inevitably take us back to the more fundamental question of human nature itself. The question which was sidestepped because of inadequate knowledge and behavioral concepts.

THE RECIPROCAL DYNAMIC AND THE STUDY OF ECONOMIC INSTITUTIONS

The reciprocal dynamic of our evolved brain structure ...the tug and pull of ego and empathy, of self-interested wealth-maximizing behavior counterpoised and complemented by empathetic, other-interested, other-maintaining behavior... provides a more accurate view of the motivational dynamics of the individual upon which any satisfactory explanation and understanding of microeconomics must ultimately rest.

The reciprocal algorithms of behavior can account for the motivational dynamic of exchange. This does not mean that they can explain all aspects of motivation. And, of course, they cannot explain or give much guidance on the cognitive issues of deciphering the environment. They can, however, take us further in the clarification of the fundamentals of social and economic exchange and they provide a superior underpinning to the study of institutions than the univariate assumption of a solely wealth-maximizing individual.

As noted in previous chapters, institutions are both political and economic in that they give order to and regulate reciprocity. The manner in which they do so is, in large part, a function of the history, traditions, ideology, and indeed the entire context, past and present, of the society. History, tradition, and ideology as conservative factors will tend to preserve whatever imbalances that exist and have existed in a society. The structured behavioral tension bound by these conservative unbalanced constraints is the ultimate motive power for the changes in hierarchy, inequality, and power...when the opportunity for change presents itself as a realizable alternative.

In addition, the social or institutional incentive structure, which is consensually acknowledged to be one of the most important factors in assessing the efficiency of institutions, will be based upon or designed consciously or implicitly to exploit or manage this innate reciprocal dynamic within the permissible constraints supported by the society, its institutions, and its ideology. In this way institutions may block, mitigate, or facilitate the increased production of goods and services in the society. Empirical research into the specific aspects of institutions that impact and channel the reciprocal dynamic should prove useful to achieving a deeper understanding of the role of institutions in the growth, stagnation, and decay of societies.

PARADIGMS AND SELF-CONFIRMING INVESTIGATIONS

The new institutional economics broadens the narrow normative paradigm of neoclassical microeconomics. In that sense it offers opportunities for a much more comprehensive and socially relevant grasp of the total question of economic exchange. The second critical point (the first being reciprocity) that has not yet been adequately grasped, however, is that new institutional economics accepts the normative desirability of the current paradigm of relentless and endless increase of productivity as an end in itself. And like the neoclassical economic approach, it operates entirely within the paradigm, without questioning the paradigm itself.

In our current and prevailing paradigm, we have created a set of institutions and an ideology that overemphasizes one side of our nature...the wealth-maximizing, self-interested. In our so-called "value-free", "positivistic", "objective" economic investigations within this blatantly normative framework, we confirm what the framework already dictates. That is: within a wealth-maximizing self-interested framework, individuals are rewarded for doing so and therefore tend to demonstrate such wealth-maximizing self-interested behavior. And when we constantly emphasize and promote the results produced within this normative framework as "objective"...as the way things really and necessarily are...we further reinforce the framework as well as the same pre-directed, predictable outcomes. It's simple and straightforward: in a wealth-maximizing, self-interested institutional framework, people do exactly that. With cooperative, altruistic exceptions, of course, that are difficult to account for.

In an alternative paradigm, we ***could*** create a set of institutions and ideology that emphasizes the other side of our nature--the empathy-maximizing, other-interested. Within that framework we could account for all exchange as empathy-maximizing, other-interested. And most of the behavior we observe in our investigations could be interpreted to fit within that motive. Of course we would then have self-interested exceptions that would be difficult or embarrassing to account for.

But the alternative paradigm would also be a distortion of our nature. Each of the two alternatives would produce behavioral tension owing to the overemphasis on one side of human nature rather than the other. In other words we would have side-effects, spillover empathetic or egoistic behaviors, that deviated from the expected and would have to be explained away...like Ptolemaic epicycles which accumulated to explain deviations in astronomy prior to the Copernican revolution.

An institutional framework which appropriately recognizes and encourages both ego and empathy, self- and other-interest, reciprocity in dynamic balance, however imperfectly it functions, moves us further in the direction of not only wealth-maximizing, but also wealth-sharing and social responsibility.

An important principle to keep in mind in any so-called scientific inquiry, but especially so in social science, is that an investigation conducted within a paradigm will inevitably tend to confirm the paradigm. Only when the discrepancies accumulate sufficiently to be glaring will an alternative paradigm begin to emerge, as in the Copernican revolution and in the move from Newtonian mechanics to relativity and quantum theory.

In the social sciences the raw materials are much more malleable than in the physical sciences. They do not represent a fixed and immutable reality. This is because we create the whole of the social sciences out of what we are. And herein lies the reason why it is so foolish and misleading to speak of "objective" social science. Society is not a given like the physical world, e.g., the laws of gravity, to be discovered essentially the same everywhere. Society has been structured, created by us, mostly unconsciously and incrementally, sometimes more self-consciously. All research within the existing created structure tends to confirm that structure since behavior operates within the constraints imposed by that structure.[39] To take the results gleaned from research within a particular ***created*** socio-institutional structure, accepted and interpreted as necessary, objective reality, does not inform us about alternatives and may limit our choices. This factor of creation indicates the limitations of Milton Friedman's axiom that..."the only relevant test of the validity of a hypothesis is comparison of its predictions with experience." (1953: 9). Adhering to such an axiom will almost certainly cause us to miss alternatives. This does not mean that, within the created institutional frameworks, we must abandon the tools and methodologies developed under the positivist pretense, but rather that we should use them with a clearer understanding of their limitations. In other words, we may use the observational and measuring tools to gather data and assess whether the institutional framework which we created is doing the job...generating the behaviors and results...that we intended or hoped that it would do.

The basic materials that we have to work with in modifying any existing social paradigm, or in creating any alternative one, are our inherited protoreptilian-mammalian brain structure topped with a massive generalizing/analyzing cortex. Out of this structure emerges the reciprocal algorithms of behavior. These algorithmic rules are the basic shaping dynamic of our social, economic, and

[39]Paul Feyerabend, a leading philosopher of science, for instance, takes the cautionary position that the meanings of both observational and theoretical terms are dependent upon the theory in which they are imbedded (1975: 320-21). Robert Heilbroner (1997) suggests that economics, as conceived today may well be a product of the capitalist system and have no applicability outside such a system. Miller goes much further when she writes: that orthodox economics "conceives a world consonant with cherished prejudice, irrespective of brute force, irreducible facts." She concludes that "orthodox economics seems to be in a permanent and unyielding state of cognitive dissonance."(1996: 97).

political lives, and the institutions and ideologies that we create, incrementally and unconsciously, or intentionally and thoughtfully, greatly channel their expression and the degree of behavioral tension that exists within our society.

It is perhaps incumbent upon us, in keeping with the focus of constitutional economics (e.g., see Buchanan 1991), to fully understand these algorithms of reciprocity and consciously design our institutions to facilitate their expression, to manage the inevitable behavioral tension of their tug and pull at all levels of social exchange, and to accomplish our normatively chosen and desired objectives...rather than leaving our institutions to fortuitous formation that may take us in directions we do not really want to go because we erroneously believe that the dynamics and features of our social world have the same inexorable objective reality as those of the physical universe.

RECIPROCITY AND THE NEW INSTITUTIONAL ECONOMICS: A CONCLUDING OBSERVATION

As noted the new institutional economics uses the same methodology of received marginalist economic theory. It adds, however, four important changes to the theory of production and exchange to accomplish the analysis of institutions. They are: methodological individualism, utility maximization, opportunistic behavior, and bounded rationality.[40]

1) Methodological individualism. The focus of analysis is changed from organizations or other collective entities (e.g., the state, society, the firm, etc.) to the individual human actor. That is, the theory of the larger social unit must, contrary to orthodox theory, now begin with and base its explanations not upon the behaving social unit, but rather on the behaving individuals within that unit.

2) Utility maximization. The individual members are assumed to seek and maximize their own self interests within the constraints established by the existing, self-interest seeking and maximizing organizational structure.

3) Opportunistic behavior. Williamson's concept of opportunistic behavior (1975) adds a further complication to self-interested utility maximization. It assumes that individuals (either principals or agents) are likely to be dishonest in the sense that they may disguise preferences, distort data, deliberately confuse issues, etc. Williamson refers to such behavior as "self-seeking with guile"or deceit.

4) Bounded rationality. In keeping with the conceptual term "bounded rationality" coined by Simon (1948; 1972), individual members do not have the perfect knowledge assumed by the orthodox theory, but have only limited ability to acquire and process information. The limitations imposed by imperfect knowledge mean in effect that not all economic exchange can be organized by contract and market but is impacted by unknown and unforeseen contingencies.

How would the acknowledgment of the algorithms of reciprocity affect these assumptions?

Let's consider the case of the free enterprise business firm. Under the new institutional economics, the self-interested utility maximization assumption, combined with the added assumption of opportunistic behavior or the seeking of

[40]For example, compare the summary in Furubotn and Richter (1991: 4-5).

self-interest with deceit or guile, seemingly sets up a unifying, consistent behavioral assumption applicable both to the firm and its individual members. Actually, in effect, the seemingly unifying assumptions pit the firm and its members against each other in an almost inevitable conflict of interests. This is because, externally, the firm is viewed basically as a self-interested and opportunistically utility-maximizing entity in the market and, internally, the positions of the individual members are, likewise, fundamentally viewed as self-interested utility- maximizing made worse by opportunistic behavior.

Nevertheless, contrary to these fundamental assumptions, the other-interested, empathetic qualities of cooperation and trustworthiness are deemed essential on the part of members to the successful and efficient workings of the firm. Cooperation and trustworthiness are the firm's necessary and desired member attributes. To overcome the assumed self-interested, maximizing behavior of the individual members in favor of cooperation and trustworthiness, which by presumption goes against the grain, imposes major production and transaction costs upon the firm. Nevertheless, the establishment of such empathetic behavior is the object of intense management concern.

Management, thus, wants cooperative, trustworthy behavior, in spite of the very paradigm that contradicts and militates against such empathetic behaviors. So management incurs great costs in attempting to overcome the implications of the paradigm by efforts to impose a standard different from the firm's market behavior on the individual members. Against the grain of the self-maximizing assumption, then, management wishes to impose a contrasting standard of empathetic cooperation and trust on the individual members. Given the overall maximizing assumption, this can only be done by rigid control, deception, or assuming that the individual members are fools. The attempt to do so, therefore, not only involves huge production and transaction costs expended on either controls or the deceptive effort, but must also contribute enormously to cynicism and alienation on the part of individual members.

The problem of double standard and conflicting messages can be avoided or mitigated by acknowledging the algorithms of reciprocal behavior based on the tug and pull of ego and empathy. Under these algorithms the firm offers compensation and benefits, an act of providing or empathy which acknowledges and affirms the ego demands of the individual member. Based upon the affirmation of ego (value, dignity), the individual reciprocates empathy in the form of cooperation and trust toward the firm and a commitment to the firm's utility maximizing behavior in the marketplace. Empathy would thus be a reciprocal factor mitigating purely self-interested, individual maximization and making more effective, less costly organization itself possible.

Under the paradigm of reciprocity, the firm could also harmonize its internal with its external operating assumptions reducing both cynicism and alienation. Under the paradigm of reciprocity, the firm would see itself as maximizing not just self-interested profit, but also empathetic provisioning done with quality and social responsibility. Of course, in keeping with the dynamics of our brain structure, the tension between ego and empathy is carried inevitably into firm's market activities in the form of tension or the tug and pull between egoistic self-interested, profit maximizing and empathetic, other-interested social responsibility. A measure of the firm's success would be how well it achieves a dynamic balance of the two.

By shifting to acknowledgment of the reciprocal algorithms of behavior, the double standard would not only be eliminated, but the market relationship would be more accurately defined...as the reciprocal relationship of exchange that it truly is...based upon our evolved brain structure.

Additionally, opportunistic behavior, especially the costly opportunism with guile, would be viewed as the exception and not as the expected behavior assumed and reinforced by the very paradigm of competitively self-interested maximization. Opportunistic behavior, as a result, would be seen not as central, but rather as deviant behavior. And it could be dealt with as such.

The problem of bounded rationality, of course, would remain.

The new institutionalist literature, following Williamson (1975), recognizes that, in the absence of bounded rationality and opportunistic behavior, all economic contracting problems would be trivial. Indeed there would be no need to study economic institutions.

There need be no fear, however, in accepting the paradigm of reciprocity. Doing so will, in fact, better define the task of the new institutional economics. It may also be expected to contribute to the reduction of production and transaction costs. It would not, on the other hand, given the persistence of bounded rationality and the need to cope with opportunistic behavior as deviancy, eliminate the need for institutional analysis or put the practitioners out of work.

As a final word of caution, even fully acknowledging and taking into account the reciprocal algorithms of behavior will not eliminate, but only help to mitigate, the costly integrative and transaction cost problems of the firm. This is because behavioral tension is inherent in any hierarchy, and despite the clearest understanding of the dynamics and best efforts of all involved, there will be an inevitable tendency to adjust the imbalances or inequities of the hierarchy, whatever specific form they take (e.g., power inequities, status inequities, economic inequities, etc.). Given the dynamic of our evolved brain structure such is the nature of the beast. It applies not only to firms but also to the administrative hierarchies of any type, economic, social, or political.[41]

[41]The inherent instability of hierarchies is indicated by the analysis of the breakdown of the Soviet administrative hierarchies (Solnick, 1998).

Chapter 14

Do All the Children Have Shoes? The Contrived Nature of Demand and Supply in Economics

In standard or received economic theory, supply and demand are the key concepts of the discipline and constitute the fundamental dynamic of the marketplace. It is recognized by some economists that, unlike the physical sciences, economics does not rest upon the constraints and laws of the physical universe, but upon a psychological dynamic. For example, Robert Heilbroner, in a short article, "The Embarrassment of Economics"(1998), expresses the perennial concern that haunts and embarrasses many economic thinkers: Is economics really a science? Heilbroner reminds students of economics, that their discipline rests not on physical constraints, but on parallel psychological stimuli...the rise and fall of prices which produces differing behavior in buyers and sellers. Heilbroner goes no further in pursuing the question of the source of these opposing parallel responses to the stimulus of prices. Ultimately, however, for a proper understanding of economics as well as human behavior, we must trace these pervasive behavioral responses back to the human brain. There is no where else to go. The answer, I contend, is that the source of these parallel behavioral responses is to be found in our evolved brain structure. That is, demand and supply are driven by the reciprocal algorithms of behavior.

The definitions of the terms ***demand*** and ***supply*** as used in economics, however, are not common sense definitions or the definitions that people use in daily social discourse. And further, they represent a truncated, not a full expression of the underlying behavioral processes upon which they rest. The definitions of economics are contrived definitions, prefabricated to produce predictable theoretical results, and to give economics the aura of being an objective science. What they actually do is create a self-confirming tautology that obscures truly significant social and political information.

The definitions given by economics to these fundamental concepts, as drawn from a widely used introductory college textbook (Heilbroner and Galbraith 1990), are as follows.

Demand consists of two components, "taste" and ability to buy. Taste, according to the authors, refers to the individual's desire for a good...and taste determines the individual's willingness to buy. The second component of demand, ability to buy, depends on possession of sufficient wealth or income (1990: 132-33). In other words, you must have the taste for something plus the ability to pay for it in order to be included within the economic concept of demand.

Based on this exclusive definition, the authors advise us with remarkable prescience that...

> The demand of the poor for food, shelter, clothing, and other necessities is small, even though their need and desire for such items may be very large (1990: 133).

The definition, then, as contrived by economics, is divorced from social reality and excludes the most critical information concerning the failure of the market to respond to social needs. It may be said that the definition excludes the ***most important index*** of market failure and social tension.

The definition of supply, although not so blatantly detached from social reality, also depends upon willingness to sell and ability to sell. In other words, the seller must be able to cover the costs of production and selling, or be able and willing to absorb the loss of supplying.

Even as presented in truncated form by economics, the algorithms of reciprocal behavior, driven by behavioral tension, come through clearly. According to the authors, supply and demand interact by their opposing behavioral dynamic (their tugging and pulling against each other) to achieve what is called an equilibrium price at the point where they intersect (1990: 136).

DO ALL THE CHILDREN HAVE SHOES?

To illustrate the working of the market dynamic of supply and demand, the authors talk about shoes. The demand for shoes declines as prices go up. That is, as prices go up, some individuals with less ability to pay, whether or not they may desperately need shoes or even clamor vociferously for them in street demonstrations, are excluded from the economic concept of demand. That is, we simply don't count them anymore. This exclusion may or may not deny or contradict the reality of the social demand or need. But it does certainly tend to hide any discrepancy from us.

On the other hand, as prices go up, the supply of shoes also goes up. More producers and sellers are willing and able to supply. At some point the declining curve of demand (as prices rise) will intersect the climbing curve of supply (as prices rise). At the point of intersection we have what economists call the equilibrium price. At this equilibrium point, economists say the market "clears". That is, suppliers have supplied shoes to all who can afford and are willing to buy them at the price suppliers are willing and can afford to supply them (1990: 136-38). A perfect market! The equation has fulfilled the tautology of its contrived definition.

But it has failed to address, even consider, the crucial social question: Do all the children have shoes?

Or do all the homeless have homes? Such is the plight of the poor. They are dropped out of the equation by the contrived economic definition of demand. From the viewpoint of economic science, we could have a perfect balance of demand and supply, a market that clears perfectly ... and one-third, half, or possibly none of the children may have shoes...or adequate food, shelter, or medical care.

THE UNSCIENTIFIC NATURE OF ECONOMIC THEORY

Economists may defend themselves from charges of callousness by saying: "We, too, empathize with the plight of the poor, but science cannot address such questions. Those are normative or value issues. Science is objective and confined by its very nature to the facts." This defense, of course, is clearly specious. The economic definition of demand was normatively contrived to begin with...not to discover and elucidate facts...but rather to fulfill a preferred (normative) tautology. And contrary to the claim and charter responsibility of science to search out the facts, the definitions of economics have the effect of excluding, not some trivial and irrelevant facts, but the most crucial facts, from discovery or investigation. As they stand the fundamentals of economics are nonscientific by their own definition.[42]

The definition given of demand, then, obscures the facts and reality of the truly important human issue. This human issue is the gap between the contrived definition of demand and the actual need or social demand. Information relating to this gap is the most important social and political information because the gap shows the social failures of the economic system...it indexes the behavioral tension ... the inequalities of the system.

This is the information that all political leaders, not bound by special interests, should need and want to know. In a nondemocratic society it tells how much force, how much coercion, how much police power, will be required to maintain the social order. In a democratic society, it tells about the tensions underlying the political process; it indexes social unrest, inadequately articulated political interests. It is information that indexes both the opportunities and the dangers confronting the society.

SOCIAL DEMAND, SUPPLY, AND EXCESS PRICE

Demand, minus the contrived limitations of the economic version of demand, may be called ***social demand***...to distinguish it from the economic definition.

To define the relationship between social demand, supply, and price, I suggest a term called ***excess price***. I would define excess price as the price at which any individual in a society who needs an item of necessity is excluded from the supply of that item.

The concept of social demand brings market exchange theory into alignment with the behavioral algorithms of our evolved brain structure. It may not simplify the task of economics, but it allows us to examine, in social and political terms, the

[42]On considering the question of the objective nature of economics, one must distinguish between the objectivity of the theory itself and the objectivity of the investigator. The two are not the same thing. Bromley observes that in policy science ... "It is ***not***(italics in original) the science -- nor the conclusions -- that are objective but rather the economist who stands between theory and individual(s) who must make a decision with economic content and implications."(1989: 233). The crucial distinction between the objectivity of the science and the scientist has often been confused in economics. See also the comments on two kinds of objectivity by Johnson (1986: 51).

successes and failures of the market. It allows us to index the behavioral tension in society, and to define our socio-political challenges and choices more clearly. Social demand aligns with social or public choice and allows us to see the fundamental unity of the disciplines and the issues they must confront.

On the side of better science, the realization that the opposing dynamics of supply and demand are driven by the behavioral tension of the reciprocal algorithms of behavior deriving from our evolved brain structure allows us to replace the semi-mystical, metaphysical, pervasive economic concept of the invisible hand, by a concept grounded more securely in evolutionary neuroscience (see Cory 1991).[43]

RECIPROCITY AND THE INVISIBLE HAND

Economist, Jerry Evensky (1993), writing on ethics and the invisible hand, notes that Smith's invisible hand metaphor represented for Smith the invisible connecting principles of the universe created by the Deity. The popular concept, then, was early and clearly based on a mystical, unexplained dynamic. Among the connecting principles, it included both self-interest and benevolence (fellow-feeling) (see Smith 1911, 1937), reflecting, perhaps, an intuited perception of the reciprocal algorithms of behavior projected by Smith on to the socio-economic system. Evensky points out that Smith's confidence in the metaphor waned over time, and in his last revision to the *Theory of Moral Sentiments* in 1789, he appealed to all citizens to put the well-being of the society first and to statesmen to construct a moral society by deed and example (Smith 1977: 319-320).

In the attempts to formalize the workings of the intuited invisible hand over the next two centuries different models were used. The rigorously mechanistic approach of Leon Walras (1834--1910), who attempted to demonstrate the equilibrium of the invisible hand, and the follow-on work of Vilfredo Pareto (1848-1923), both, of the so-called Lausanne School, aped the model of Newtonian mechanics (Ingrao and Israel 1990: 87-138). The mechanical model later gave way to the mathematical modeling of John von Neumann, Oskar Morgenstern (von Neumann and Morgenstern 1944), and Paul Samuelson (1947), which was inspired mainly by developments in modern physics. This led, in turn, to the rigorously axiomatic approach of Gerard Debreu and Kenneth Arrow (1954).

The pursuit of the question of a market equilibrium produced by the intuited invisible hand has been formally divided into three categories of existence, uniqueness, and stability. The axiomatic approach of Debreu and Arrow finally achieved somewhat satisfactory results with regard to the question of existence. That is, they demonstrated formally the existence of general economic equilibrium under very general assumptions inherently fundamental to the basic Walrasian theory. On the problems of uniqueness and global stability, however, the efforts have been disappointingly unsuccessful. It seems perhaps that a new paradigmatic approach may be necessary (Ingrao and Israel 1990: 362).[44]

[43] Cosmides and Tooby(1994) also suggest that the invisible hand is structured by the interaction of human minds, or brains.

[44] An alternative paradigm of hermeneutics has been proposed by Addleson (1995). But this paradigm would require the abandonment of any search for equilibrium.

It is safe to say that basing the paradigm on explicit or implicit models of classical mechanics, followed by models inspired by modern physics, was inappropriate. Market equilibrium is not driven by gravity or any other laws or principles of physics. Market dynamics are the product, sometimes highly rarefied and extended, of the social exchange activity of human beings. The reciprocal dynamic of our evolved brain structure, the dynamic probably intuited by Adam Smith as the hand of Deity, truncated in order to depersonalize ego and empathy and eliminate change and development in the manner of the Arrow-Debreu formalization, does suggest, and can certainly include, the existence of a very general economic equilibrium. Whether the algorithms of reciprocal behavior, with their inherent genetic, experiential, behavioral, and statistical variability can be adequately formalized to deal with the problems of uniqueness and stability, however, is a question that must be left open to further research.

Uniqueness and stability are especially problematic in that they are categories that proceed from implicit mechanical or physical images, lacking the complexity of a biological dynamic. The current general equilibrium theory is based upon a static set of assumptions that eliminate factors of change, growth, or development in order to try to get to a unique and stable general equilibrium. Such an equilibrium, however, is incompatible with the full biological brain dynamic. Any equilibrium achieved within the biological paradigm would necessarily be a dynamic one, which may be better described as an ***equilibration*** (after Piaget 1977: 4-38, as rooted in Hegel). An equilibration is a constantly changing equilibrium which never returns to the same place but tends to move progressively from the former equilibrium to a new equilibration, which is inclusive of the former equilibrated state plus change. To achieve a proper demonstration of existence within this biological (rather than mechanical or physical) model, the static constraints that were necessary to demonstrate general equilibrium would have to be relaxed or abandoned. Given the dynamic, shaping, and creative nature of the brain dynamic, the challenge to demonstrate the existence of equilibrations, by adding growth, development, and change to the concept of equilibrium, may itself prove to be indeed challenging. On the other hand, under the biological model such equilibrations may never be unique or stable in the static sense, and therefore, being impossible to demonstrate, uniqueness and stability may prove to be meaningless categories of the formal problem. There is, nevertheless, the probable obligation, if not necessity, to pursue the indicated research program.

THE INVISIBLE HAND AND THE UMBRELLA OF SOCIAL EXCHANGE

As noted in chapter 7, studies of exchange in primitive societies tend to show that exchange rests not primarily on what are assumed to be economic but rather on social factors. The overriding fact is that humans socialize. They are completely immersed in a social context. They are mammals and the social glue emerges from the bonding mechanisms of the mammalian affectional programming. These mechanisms potentially conflict with the earlier protoreptilian programming from which ego and self-interest emerge. With the elaboration of the neocortex, bringing generalization and language, the algorithms of reciprocal behavior emerge as the dynamic basis of social exchange.

There are many forms of social exchange. Individuals and groups associate together, dance and sing together, exchange information to include trivial gossip, gather to keep each other warm, participate in mutual protection, in sex, and in mutual nurturing and affection. All such activities are activities of social exchange in which give and take, ego and empathy, self- and other interest are engaged. ***Economic exchange exists as a subset within the overall umbrella set of social exchange.***

In the small, largely extended family units of primitive foraging societies, behavioral tension tended to keep significant inequalities of resources and power from developing. It seems that the reciprocal dynamic, the invisible hand of brain structure, which evolved as an adaptation within that same foraging social structure, functioned within the range approximating dynamic balance. The tension of imbalance was personally, subjectively felt, adjusted, and objectively manifested in the generally egalitarian structure of such small, kinship-based foraging societies. With the shift to larger, less personal political units, which probably accompanied or brought about the shift from gift to transaction and the beginning of the market, the invisible hand of brain structure likely began to function less reliably. Larger, less personal, political units went hand in hand with division of labor and differential accumulation of resources, then wealth and power. Under such circumstances, coercion and force are increasingly required to control the accumulating behavioral tension that strives toward a correction of the inequities. Nevertheless the tension is always there, acting as an invisible hand pressing for adjustment and change during breakdowns of power and when there exists opportunity, perceived or real, for success.

The invisible hand of the reciprocal algorithms works well only in societies with institutions that facilitate, rather than impede, the shifting balance. Such institutions must facilitate freedom of social (to include economic) exchange in all its forms. Political participation is also a form of social exchange. Whatever obstructs full and free political exchange, whether powerful interest groups or institutions that protect wealth accumulation, creates behavioral tension and inequality. For the invisible hand to work effectively in larger societies, it requires the intervention or facilitation of what may be called an ***intentional*** hand as an institution, or set of institutions, to prevent the excessive accumulation of wealth and power and to relieve the associated, accumulated behavioral tension. As economic philosopher Robert Nozick has observed "...an invisible hand process might better arise or be maintained through conscious intervention."(1994: 314). Freedom and equality are often seen as conflicting ideals. Actually, they need not be. They come into conflict only when freedom is defined as the freedom to alter the capacity for the free exchange by others by creating inequalities of wealth and power in the exchange process.

The full realization of the intentional, maintained invisible hand of full and free social reciprocity would point toward a utopian society in which freedom and equality do not conflict. This would produce a thoroughly democratic communitarian (not communist) society in which all could engage in social exchange freely and equally and in which the right and capability to so engage would be protected by institutions which prohibited the accumulation of the wealth and power that inhibit the free social exchange of others.

This is indeed an ideal. One that we may approach but never fully attain because of the necessity to incent for productivity and/or protection. Nevertheless, this is the direction in which the reciprocal algorithms of our evolved brain structure ultimately beckon, if not lead us.

Chapter 15

The Reciprocal Equation in Behavior, Social, and Economic Exchange: An Interim Summing Up

The pervasive influence of the dynamic of the reciprocal algorithms of behavior can be understood when one realizes the relationship between behavioral tension, costs, production, and exchange. Understanding this can go a long way toward unifying the social sciences

As an interim summing up, let's start from a simple exchange of gifts between two parties.

An act of giving or, in economic terms, supplying, is an act of empathy. It creates behavioral tension (cost) in the giver. In other words, it costs the giver something to give the gift. Internally to the giver, she/he expects a reciprocal offset in payment to ego. On the other hand, the giving or supplying creates behavioral tension in the other party. This tension is manifested as a necessary urge or obligation to return a gift or service of equal cost. When the second party does so, the reciprocal exchange is complete. If approximately balanced, there is no residual behavioral tension related to the exchange.

As we increase the behavioral tension (costs) of the empathetic gift by giving something that requires more effort or value to produce or obtain, we accordingly increase the reciprocal necessary to offset the gift.

The same dynamic holds when we look at exchange from the market level. Specialization, resources, and organization (hierarchy) for production, transaction costs in the exchange process, discussed in preceding chapters, increase the behavioral tension, the costs, necessary to make the act of giving or supplying. And this added behavioral tension or cost requires a directly proportional increased return payment to offset it.

RECIPROCAL EQUATIONS OF SOCIAL/ECONOMIC EXCHANGE

Simply put, these are the reciprocal relationships and their variant expressions expressed as balanced equations. The balance, because of the nature of the dynamic, is always a dynamic one. To keep the relationships clear, I have reversed the usual English phrasing, where appropriate, for the variables. That is:

Ego is expressed by take, demand, price.
Empathy is expressed by give, supply, cost.

Ego = empathy	E = Em	E/Em	or	Em/E	= 1
Take = Give	T = G	G/T	or	T/G	= 1
Demand = Supply	D = S	D/S	or	S/D	= 1
Price = Cost	C = P	C/P	or	P/C	= 1

COST/BENEFIT ANALYSIS AND THE EQUATION OF INTERNAL RECIPROCITY

In economics cost/benefit analysis or benefit/cost analysis (cf. Mishen 1988, Schmid 1989) is the application of the theory of collective action applied to issues of public finance where measures of profit, the excess of gains over costs are not clearly applicable. CBA seeks to achieve measures of a public good vs. the cost to the public. In a sense, CBA measures effects internal to a single (albeit collective) actor...the public. When so applied, internally to a single actor, whether the collective actor of the public or to an individual person, CBA does not concern the exchange relationship between the affected actor and another party, except perhaps indirectly. In other words, we are concerned with the benefits received by one actor in comparison with the costs to that same actor. When applied to the indivisible, individual person, this is the analogue of the tug and pull of ego and empathy within a single individual. It is directly governed by behavioral tension.

The match with the fundamental formula of reciprocity between actors, expressed in the previous section, is somewhat obscured by two factors: 1) the limit of focus to one party to the exchange, and, 2) the shifting of definitions from the price/cost relationship above. Depending on preferred perspective, cost can be seen either as egoistic or empathetic, and benefit may be seen as either/or, likewise. But the reciprocal relationship is nevertheless there. When costs and benefits equal, there is balanced internal reciprocity; one gets the equivalent of what one gives. Neither ego nor empathy suffers. Behavioral tension is dynamically balanced. Residual behavioral tension is essentially lacking or minimal.

Benefit = Cost B = C B/C or C/B = 1

THE ENCOMPASSING RATIO

The ratio between the two variables indexes the degree of behavioral tension, inequality, hierarchy, domination/submission (potential or actual), and/or, power. This is the fundamental dynamic of social exchange. Complex market exchange builds upon, extends, and ramifies this fundamental relationship.

The difficulty of determining the various weights of the variables to be applied in each use of the equation is complicated by cognitive limitations, perception, information, measuring; and, in contractual exchange, enforcement. Such factors complicate the relationship and the ratios but do not alter them fundamentally.

BEHAVIORAL TENSION AS A FUNCTION OF RECIPROCAL IMBALANCE

Behavioral tension is a function of reciprocal imbalance and may be expressed by the equation

$$BT = Em/E$$

Since behavioral tension, ego, and empathy, as magnitudes, evade precise numerical quantification, approximate values must be assigned. If we allow a range of 1 to 10 for the variables Em and E, and assign the value of 8 for empathy and 4 for ego in a particular social exchange, we can express the equation as follows:

$$BT = 8/4 = 2$$

In this exchange, then, there is a Behavioral Tension factor of 2. Given the nature of the dynamic, this means that one party to the exchange has given twice as much as he/she received and the second party has taken twice as much as she/he has given. This has created a tense relationship of imbalance, which may be expressed in various ways as a debt/debtor relationship, credit/creditor, inequality, dominant/submissive, hierarchical, or a relationship of power inequality.

In cold and pure math, the single exchange stands alone and is complete in itself with no future or after effect. But given the human dynamic, the relentless and ongoing tug and pull of ego and empathy, combined with cognitive capacity and memory, the effects are long-term and cumulative.

A one-time exchange will be remembered; the behavioral tension retained, to be rectified, if possible, in a future transaction.

In a continuing exchange relationship, reciprocity may be approximately or dynamically balanced out in a series of exchanges over time.

If, however, such an unbalanced exchange relationship continues over time, the behavioral tension, the dominant/submissive character of the relationship, the inequalities, the hierarchy, the power differentials, all become intensified and cumulative. The cumulative behavioral tension stresses the exchange relationship and provides the dynamic which drives the process of change, peaceful and evolutionary, or violent and revolutionary, depending on the ratio of cumulative intensity and the perceptions of the realistic possibility of change.[45]

BEHAVIORAL TENSION = SOCIO-ECONOMIC TENSION

The same relationships maintain in our economic equations of demand/supply, price/cost, cost/benefit. Economic science attempts to see these exchanges as objective, mathematical, and without subjective components. This is illusory, self-deceptive, and potentially very dangerous. Behavioral tension is always present, never absent, and is cumulative unless mitigated.

The same equations pertain in our domestic as well as global trade relations. In our domestic situation most of our social ills and discontent reflect the ongoing and

[45]Compare Sen 1997: 1. Sen's well-known work *On Economic Inequality* (first published in 1973 and expanded in 1997), which analyzes the various attempts to define and standardize measures of inequality for policy purposes, illustrates the great difficulty in achieving objective measures of the highly subjective perception of inequality.

cumulative effects of unbalanced reciprocity in exchange. The behavioro-socio-politico-economic tension underlies the calls for social equality, equality of opportunity, nondiscrimination, and justice. On the global side, this is clearly evidenced by our ongoing tensions with one of our major trading partners, Japan. The ongoing, perceived deficit relation is packed with the behavioral tension from unbalanced reciprocity. This tension has even spawned exaggerated premonitions of a coming war. In our policy-making and policy rhetoric, we hide behind an ideological facade of free and open exchange, free and open trade...but what we also want is a minimum relationship of balanced reciprocity. Such is the nature of the evolved brain dynamic. We will accept a position of advantage, but empathy allows us to settle willingly, and without too much discomfort, for a minimum condition of balanced reciprocity.

When the fundamental dynamics of our triune reciprocal brain structure are fully grasped and appreciated, they provide a giant step toward unifying the social sciences as well as connecting them with neuroscience. The prospect calls to mind the words of Kenneth Boulding written in the preface to his 1950 book, *A Reconstruction of Economics:*

> I have been gradually coming under the conviction, disturbing for a professional theorist, that there is no such thing as economics – there is only social science applied to economics (1950: vii).

Chapter 16

The Culture Bound Nature of American Economic Theory

In preceding chapters, I have presented arguments to show that received economic theory is not objective science, but a normative, humanly-created paradigm that shapes results in its own pre-contrived image. Three factors discussed thus far undercut the claim of received economics to objective, scientific status. They may be summarized as follows:

First, unlike the physical sciences, economics is not based upon universal, immutable laws like gravity and the regularities of chemical interactions. Economic systems are socially created by the human brain interacting with like brains under physical as well as imposed, self-created constraints. We can choose to change our social systems. We cannot choose to change the physical laws of the universe.

Secondly, received economics not only erroneously conceives the nature of humans as solely wealth-maximizing, self-interested creatures, it misses the glaringly evident fact that all exchange depends upon reciprocity. The primary vocabulary of economics... such terms as, exchange, transaction, transaction costs, demand and supply... are terms of reciprocity. Exchange, whether viewed narrowly as economic, or more broadly as social, is simply, straightforwardly, and necessarily a reciprocal process. Further, there could be no exchange or transactions...no market of any kind...based on self-interest alone. Empathy, the capacity to recognize and participate in the self-interest of others, is the necessary balancing reciprocal to self-interest. Without empathy we could not engage in exchange. We would not know what to offer, or how to offer it. Without empathy we would have no way to respond effectively to another's needs, to effectively fulfill the supply or provisioning side of the fundamental economic equation.

Thirdly, the most basic terms of economics, demand and supply, are defined arbitrarily to accommodate the self-fulfilling prophesies of the normatively contrived economic paradigm. The definition of demand, especially, excludes, rather than reveals, what are probably the most relevant social and political facts... and the most conspicuous failures of the market system.

Hampden-Turner and Trompenaars catch the essence of the Anglo-American bias when they write about the self-interested orthodoxy of neoclassical economics.

> Adam Smith's famous doctrine of self-seeking at the roots of Homo Economicus is perhaps the world's leading example of cultural bias and historical circumstance disguised as a principle of science. (1993: 53).

Of course, it can equally be argued that the more empathetic communitarian emphasis of Japan, France, and Germany may also be accurately considered examples of cultural bias and historical circumstance. The only difference is that these cultures have not so conspicuously and widely promoted and publicized their cultural biases and historical circumstances as "science".

As the world gropes for more understanding and an economic system that works well for the new global realities, a full appreciation of the reciprocal nature of social exchange may emerge and serve to achieve a closer alignment of values and practices within the community of capitalistic, free-enterprise nations.

Received economic theory is not a universal science, but rather a culture-bound product of primarily the Anglo-American cultural perspective. This perspective has prevailed to this point because of America's economic dominance and the presence of a politically polarized global circumstance. The world has changed, and the historical circumstances have changed as well. It is inevitable that economic thinking will accordingly be modified to fit better with the new realities. The alternative cultures have lacked the credibility, the prestige, if not the interest, to present their case articulately and convincingly. This alternative presentation needs to be done. Free-enterprise capitalism can overcome its defects and be worthy of becoming a global, normative system only if the alternatives are fully explored, debated, and tested.

Chapter 17

Public Choice Theory and Political Science

In recent decades the studies of political science and economics have largely gone their separate ways. The emerging new emphasis on political economy reflects a change in that state of separation. It is based upon the perceived underlying unity of the political and economic sides of social life. Political economy emerges at the point on a spectrum at which the disciplines of political science and economics, coming from differing perspectives on social life, flow into each other or converge. At this point there is occurring an exchange or blending of theory and methodology.

Indicative of this convergence is the importing of self-interested rational choice theory into political science under the rubric of public choice. Despite the advantages of uniform theory and methods bridging the two disciplines, there is also a downside. The preexisting problems and distortions are carried from economics into the public choice literature based upon the self-interest emphasis and the failure to recognize the reciprocal nature of all exchange and choice. A paragraph excerpted from an introductory text on public choice is sufficient to illustrate this claim.

> Despite the absence of a human or digital "Big Brother"(quotes in original), chaos and anarchy do not exist in the economy. The daily -- and generally detailed, arduous, and boring--tasks of feeding, clothing, housing, transporting, and entertaining the population are accomplished without fanfare or centralized control. The source of this order is individual ***self interest*** (emphasis mine) channeled by competitive supply and demand incentives. No individual spends time contemplating the effects of his or her actions on the total welfare of society. Rather, we individuals make decisions to improve our own welfare, but, in the process, our decisions benefit other individuals in society. A farmer considering the use of his 100 acres does not calculate the total quantities of various products required by the residents of a neighboring city, nor does he estimate the quantities that are likely to be supplied by other farmers. Instead, he looks at prices and costs and asks, "Will I increase my welfare more by planting corn, wheat, or soybeans?" (quotes in original) (Johnson 1991: 53-4).

This paragraph (and similar ones can be selected from almost any text or monograph), with its emphasis on self-interest, obscures the fact that the market operates in a pervasive social context and the ordering principle is not self-interest but is, instead, reciprocity. There is an entire community, nation, or world out there stating its egoistic demands or wants, which are indexed by the price mechanism. The farmer, or any other businessperson, looks at this egoistic demand or want indicator of price and decides how much or what empathetic services or products she/he will produce or perform for provisioning or responding to this demand. The farmer or businessperson then fulfills the empathetic, nurturing, or provisioning side of the economic equation, fulfilling the egoistic demands of the individuals grouped into the concept of market. She/he then receives payment in reciprocal acknowledgment to his/her own egoistic demands or needs...and the reciprocity is complete. The mechanism is reciprocal because if the community, nation, or world were not out there, the farmer or businessperson would simply not provide.

True, the market described here is impersonal and rarefied. Therefore subjective feelings of ego and empathy may be weak to almost nonexistent. Nevertheless, the reciprocal algorithms of behavior are clearly the driving source of the market mechanism...not self interest alone as claimed by the prevailing market theory and set out as gospel by economic, rational choice, and public choice texts. Market or exchange theory, as it exists today is inaccurate, distorted, and has the unfortunate side effect of promoting self-interested egoism, Social Darwinistic propensities, and cynicism throughout the society, by failing to understand and grasp the empathetic nature of the provisioning required to fulfill the equation.

In political science one of the most important issues is that of legitimacy...the question of how people are bound into a political community...a local, national, or even world-wide identity. Patriotism and loyalty illustrate that empathy, with its roots in affection, accompanies identification with the political unit, community, nation, or perhaps world, as kin, as benefactor, as provider of safety and services. When the political unit provides, the citizens reciprocate empathy, attachment, identity, loyalty. It would be easy to obscure these essential issues of politics within a self-interested, rational choice model.

FROM CONFLICT TO RECIPROCITY: THE IMPLICATIONS FOR NEW INSTITUTIONS FOR A NEW MILLENNIUM

The reciprocal algorithms of behavior prescribed by our evolved brain structure show us how to get from conflict to reciprocity by the balancing of our self-interested and empathetic programming. These algorithms of reciprocity have been shown to be the guiding dynamic of social organization and exchange. What then do they require of us in building new institutions and behaviors to accommodate self-consciously, in a more enlightened manner appropriate to the new millennium, their inevitable dynamic?

First, I would point out the implications for orienting the discipline of political science. Although it has stirred controversy and cleavages, there is no question that the rational choice model of economics and exchange theory is becoming the dominant model in political science (Bates 1997, Johnson 1997, Lusick, 1997). By adopting the alternative model presented here, political science connects with neuroscience, as well as avoiding the negatives, the moral discomfort and criticism that have plagued the overly self-interested economic and exchange theory models. It also gains a heuristic that can better account for such important political phenomena as loyalty, commitment, and the shifting involvements of private interest and public action (e.g., Hirschman 1982). Finally the model shows the moral basis of exchange and choice which is increasingly important to our global society, and avoids the implicit and troubling academic indorsement and propagation of a one-sided self-interested egoism in public affairs. And hopefully, with the dethroning of unmitigated self-interest by the appropriate acknowledgment of the balancing role of empathy, the last vestiges of Social Darwinism will begin to fade from our social, economic, and political thought.

As the basis of our political thinking, we might acknowledge that the pursuit of self-interest without the mitigating engagement of empathy is an inevitable state of war...the war of all against all seen as the basis of exclusively self-interested political society by Thomas Hobbes in the 17th century. Contrary to Hobbes and others who have followed him, our evolved brain structure clearly shows that although humans may perhaps function for a time on a basis of exclusive self-interest, this is not the natural state of humankind. It is a distorted caricature upon which no lasting society has ever been built from primitive to modern times. There is a growing realization in economics that an economy made up of atomistic, wealth-maximizing self-interested individuals would have insufficient structure to survive. This point was made by economist and recent Nobel laureate Amartya Sen (1979) when he argued that economic man had too little structure. More recently, Fabrizio Coricelli and Giovanni Dosi tell us that most recent accounts of the overall economic order reveal a "curious paradox." Although such accounts begin with a faith in both the invisible hand and the capacity of the individual actor to process information accurately and make choices freely, they conclude with outcomes showing "a very crippled hand, incapable of orderly coordination even in extremely simple environments." (1988:136)

And even Kenneth Arrow, Nobel prize recipient and the acknowledged creator of social choice theory, has commented in an oral history interview:

> People just do not maximize on a selfish basis every minute. In fact, the system would not work if they did. A consequence of that hypothesis would be the end of organized society as we know it (Arrow in Feiwel 1987: 233)[46]

And economist Thrainn Eggertsson of the University of Iceland refreshingly and candidly puts it in terms that match those used here when he writes: "A society where everybody behaves solely in an ***egotistical*** and ***cold-blooded*** (emphasis mine) fashion is not viable." (1990: 75). It is further interesting that Eggertsson, intuitively, as most of us do since we all share similar brain structures, associates egotistical with cold-bloodedness...our protoreptilian or premammalian heritage. Our mammalian heritage gives us warm-bloodedness, nurturing, the capacity for cooperation, trust ... and the warmth and comfort of friendship.

Now that we can more clearly see the natural state of humankind, proceeding from our evolved brain structure, we can proceed to more self-consciously and intentionally construct our institutions and conduct our behaviors within these institutions to exploit these insights.

Concerning our domestic institutions and behavior, the reciprocal algorithms of behavior driven by behavioral tension, as the foundation of all social organization,

[46]This is increasingly recognized in the new institutional economic literature although no effective model to account for it has yet been devised. For other examples, see also Raymond Plant's article on the moral limits of markets (1992) and Lars Udehn (1996: esp., pp.60-114.)

prescribe that such institutions in their structure and methods of operation should facilitate the give and take, the reciprocity, among the citizenry, that allows the expression of the tension between self- and other-interest as it tends toward balance.[47] Such facilitation should begin early, especially in our socially critical educational institutions, and it should permeate the socializing curricula throughout the education process.

This means such institutions must respect and facilitate both dynamic aspects of our makeup. They must preserve and facilitate the freedom to express our ego, our self-interest, our individuality, in the form of self-expression, productivity, and creativity. And they must also facilitate and cultivate our expressions of empathy, relatedness, and social responsibility. The proper balance of the two means that individual self-interested creativity and productivity is performed empathetically in the process of social exchange as a gift or contribution to society. Creativity and productivity in fact make no sense except as a gift or contribution to be shared with others...with society. Said another way individual creativity and productivity invariably have a social context.

This tug and pull between ego and empathy, self-preservation and affection, is the biological source of the tensions between our values of liberty vs. equality that so pervade our modern thinking. The innate dynamic, that we all share and strive to articulate, divides well-intentioned scholars into opposing camps depending on which derived value, liberty or equality, we feel most strongly about, and our assessment of the means by which we can best achieve the most socially desirable balance between the two.

Although the discussion at the end of chapter 15 indicates that in a highly idealized, democratic communitarian society, liberty and equality need not theoretically be in conflict, this is an issue with deep divisions in current thinking. Total liberty would seem inevitably, even on an ideal level playing field, to involve some inequality, because of innate individual differences in intelligence, skills, talents, energy, and/or developmental factors of health, age. Total equality would likewise, seem inevitably to involve some suppression of liberty, because it would, inhibit or redistribute the advantages gained through the same innate and developmental factors. As radical institutionalist William Dugger, who exhibits a frank, emotionally-charged, and clearly normative position on inequality, remarks: "We can pretend to be value neutral about inequality, but we never are."(1996:21). On the other hand, Friedrich Hayek worries normatively, and with evident emotion, about social or distributive justice when he writes:

> I am not sure that the concept has a definite meaning even in a centrally directed economy, or that in such a system people would ever agree on what distribution is just. I am certain, however, that nothing has done so much to destroy the juridical safeguards of individual freedom as the striving after the mirage of social justice. (1991: 388-89).

Such are the normative variants that our reciprocal brain structure pushes upon us, as ego and empathy, self- and other-interest tug and pull against each other

[47]The findings from brain structure, gene-theory, and ethology reported in this book support the new emphasis on empathy, cooperation, and altruism by scholars in economics, social and political science (e.g., Mansbridge 1990; Etzioni 1988; Frank 1988; Hirschman 1982; Margolis 1982).

within our skulls and between us in society. And such clearly normative, value-bound variants are not dispassionate, but are emotionally-charged as is equally clearly displayed in the rhetoric of both Dugger and Hayek. The value as well as the emotional charge reflects the behavioral tension that drives the dynamic of the reciprocal brain.

The challenge that we face in creating new institutions is not the irreconcilable dilemma of choosing to reinforce one extreme or the other, but rather to achieve a more reasonable (less tension-filled) balance between the two essentially pragmatically conflicting motivations and their externalized social values. It is my normative position, desire, and hope, that we can do a lot better job of this balancing act. The dynamic of behavioral tension also tells us that these institutions and methods we create must be flexible since the balance will by its very nature be a constantly shifting dynamic balance as self- and other-interest, the dynamic forces of behavior, tug and pull against each other in daily interplay within the context of institutions at all levels of society. One of the most difficult questions will be how do we maintain openness and flexibility within the bureaucratic institutions that almost inevitably result from collective action to facilitate social exchange and a more even distribution? Yet openness and flexibility will be essential since rigidity and hierarchy will impede the accommodation of behavioral tension in a shifting balance and will build unrest within the society.

In our global relations, given the increasing level of social exchange brought on by the information age, the same dynamic and the same guiding principles will apply. In our trade and power relationships, the institutions must be designed to facilitate, and in some cases to adjudicate, the shifting, dynamic tendency toward reciprocal balance. To the extent that the institutions and behavior fail to accommodate and facilitate these dynamic processes, international relationships will be marred by the accumulation of behavioral tension and the threat of economic and armed conflict.[48]

How best to structure the new institutions will be a matter for careful consideration. The structure of any particular institution or organization, as well as its methods, may well be situational and specific. But always, however, the structure and methods must be flexible and open to ready change. The particular structures and methods, then, may vary as needed. Given the dynamic of our evolved brain structure, the facilitation and active promotion of a dynamically-balanced reciprocity is the only viable long-term social, economic, and political principle upon which we can build a new global society in the coming millennium.

[48]The findings also support the idealist school of thought in international relations in which empathy and cooperation are emphasized over the singular emphasis on power, or egoism of the realist school which has been dominant since World War II. Idealism, as the second major school of thought, is now making a resurgence (e.g., see Keohane 1990; Steiner forthcoming; Beitz, 1979; Inglehart and Rubier 1978). For a discussion of reciprocity in international relations, see Keohane (1989: 132-57).

Chapter 18

Conclusion

As renowned Spanish neuroscientist and Nobel prize winner, Ramon y Cajal, reportedly said: "We can never understand our universe until we understand the human brain which created that universe." This may, or may not, be overstating the case somewhat, but the reciprocal brain is, indeed, the dynamic, shaping mechanism across the interdisciplinary spectrum from evolutionary neuroscience through the alternative social perspectives of anthropology, sociology, economics and political science. The combining of an earlier protoreptilian, self-preserving tissue complex with a mammalian nurturing, other-preserving tissue complex, overlaid by a massive, generalizing/analyzing cortex, which adds rational, cognitive capacity combined with language, defines our essential humanity.

From the dynamic of this triune modular brain derives the reciprocal algorithms of behavior, the tug and pull of self-preservation and preservation of the species, ego and empathy, self- and other-interest. These algorithms are the shaping dynamic of human social organization and permeate that organization from all perspectives ...anthropological, sociological, economic, and political. Their shaping influence is absolutely pervasive.

In anthropology and sociology, the algorithms are expressed in the ubiquitous norm of reciprocity, which has been almost overwhelmingly empirically observed and reported to be the operating principle in the social exchange of all societies from primitive through modern. They underlie the premarket expressions of exchange from simple gift-giving and barter, to the foundations of contractual exchange, to the highly rarefied dynamics of modern free-market exchange. In interaction with the differential endowments of individuals and groups of individuals, as augmented by social institutions and ideologies, they shape the operation of power-relationships, and hierarchy, which are in the final analysis forms of social exchange.

This evolutionary dynamic of our three-part brain, motivates the reciprocity, the give-and-take, at all levels of our interactive, social lives. Primitive humans intuited this well, or else functioned automatically, and lived according to its promptings. Although they did not understand it as a function of our evolved brain structure, anthropologists such as Mauss, Malinowski, and Levi-Strauss have shown us that whole systems of social interaction were built upon this principle of reciprocity.

Our human brain structure adds the bonding glue of mammalian affection to the independent self-preservation of ancestral protoreptilian life forms in a state of tension that demands, at our level of evolved consciousness and capacity to generalize, the reciprocity of behavior that makes possible the evolution and maintenance of society. The universal norm of reciprocity, then, is grounded in and expresses the evolved bioneurological structure.

It is worth noting that in spite of the millions of years during which reptiles dominated the earth, they never developed a society or civilization of any kind. They lacked the necessary features of the brain and nervous system that would permit interactive social bonding and exchange. Society, as we know it, which is based upon reciprocity in social interaction, had to await the appearance of mammals...and after that, the coming of humankind.

In emphasizing the basic role of brain structure in shaping the social, economic, and political aspects of our society, I do not mean to deny the importance of social history, social evolution, the evolution and dynamics of institutions, and the growth of technology. To understand where we are, how we got here, and how we can shape our future better, we need to know everything we can about all these things, how they interact, and what short and long term social effects they produce. And we must often learn by trial and error, pragmatically, by testing which institutions and practices will work to meet our consciously chosen social objectives, and which will not. We must not become bound by ideology or coercive power, but always keep our options open.

The preservation of self and species is the imperative of the human brain. Unlike the physical sciences, our social sciences are not bound and governed by fixed laws and relationships. Our social sciences, then, can never be scientific and objective in the sense of the physical sciences. Our societies, our social sciences, are created by us. The illusion that they are fixed and inevitable is a dangerous and very limiting illusion. The only regularities we get in the social sciences, with the exception of the flexible reciprocal brain dynamic expressed varyingly under physical or imposed constraints, are those produced by the particular paradigms that we choose to operate within and accept as truth or reality...such as the obviously limited and constrained regularities of classical economic theory. Classical economic theory is merely a contrived alternative which operates rather poorly even in a limited capitalist market paradigm. And it obscures other alternatives when it makes a false claim to be the only scientific, objective alternative form of a freely operating system of social exchange. We may do better to think of the society as a whole, the social sciences as one, and focus upon the wider issue of social exchange, rather then the narrower focus of economic or political exchange. The dynamic balancing of the reciprocal algorithms of our brain structure will be expressed and accomplished across the entire spectrum of social exchange, not in any narrowly defined, artificially delimited segment of the total process.

More practically, the conclusions of this book indicate that beyond increasing development and productivity within carefully monitored environmental constraints, to include the proper relation between population growth and technological development to extend the carrying capacity of the environment, the guiding principle for new socio-politico-economic institutions in both their structure and behavior should be that they are designed and operated to facilitate the free expression of ego and empathy, self- and other-interest, as they tend toward a dynamically balanced reciprocity in all areas of our social, economic, and political lives.

This guiding principle of dynamically balanced reciprocity must likewise be extended to our international trade and power relationships. The logic of these algorithms, deriving from our evolved brain structure, reveals that anything less than balanced reciprocity is a source of tension and potential conflict both within and among nations. Unbalanced reciprocity, whether caused by the institutions and practices of the dominant or subordinate trading partner, produces the well-known tension-laden negatives of exploitation and imperialism in global affairs. In the emerging global society, the cooperation, the coordination, the adjustments and accommodations necessary for our survival into the future can never be formulated on the basis of self-interest alone. Our self-interest, whether expressed as local, state, national, or regional interests, must be tempered with empathy for all others who share the planet with us. The logic of our brain structure tells us clearly, and we all know it intuitively...that while each of us may seek advantages for ourselves, our extended groups... none of us, not one of us, will knowingly and willingly settle for less than essential equality on all significant issues of survival and humanity. Empathy is what allows us, willingly and caringly, to make this necessary accommodation to live in security, peace, and shared prosperity with others.

Bibliography

Addleson, Mark. 1995. *Equilibrium Versus Understanding*. New York: Routledge.

Alexander, Richard D. 1987. *The Biology of Moral Systems*. Hawthorne, N.Y.:Aldine de Gruyter.

Aldridge, J. Wayne, Kent C. Berridge, Mark Herman, and Lee Zimmer. 1993. "Neuronal Coding of Serial Order: Syntax of Grooming in the Neostriatum." Pp. 391-95 in *Psychological Science*. V. 4, Nr. 6 (Nov).

Allsopp, Vicky. 1995. *Understanding Economics*. London: Routledge.

Alt, James E. and Kenneth A Shepsle, (eds). 1990. *Perspectives of Positive Political Economy*. Cambridge: Cambridge University Press.

Altman, Joseph and Shirley A. Bayer. 1997. *Development of the Cerebellar System: In Relation to Its Evolution, Structure, and Functions*. New York: CRC Press

Arndt, Helmut. 1984. *Economic Theory VS Economic Reality*. Trans. by W. A. Kirby. Michigan State University Press.

Arrow, Kenneth J. 1987. in *Arrow and the Ascent of Modern Economic Theory*. Ed. by Feiwel, George R., London: Macmillan Press.

Arrow, Kenneth J. 1963[1951]. *Social Choice and Individual Values*. New Haven: Yale University Press.

Arrow, Kenneth J, and Gerard Debreu 1954. "Existence of a Equilibrium for a Competitive Economy." Pp. 265-90 in *Econometrica* 22.

Arrow, Kenneth J. and F. H. Hahn 1971. *General Competitive Analysis*. New York: North-Holland Publishing.

Arthur, Wallace. 1997. *The Origin of Animal Body Plans: A Study in Evolutionary Developmental Biology*. Cambridge: Cambridge University Press.

Austin-Broos, Diane. 1996. "Morality and the Culture of the Market." Pp. 173-83 in *Economics and Ethics*. Ed. by Peter Groenewegen. London: Routledge.

Axelrod, R. and W. Hamilton. 1981. "The Evolution of Cooperation." Pp.1390 in *Science*.V. 211.

Baal, J. van . 1975. *Reciprocity and the Position of Women*. Amsterdam: Van Gorcum, Assen.

Baars, Bernard J. 1997. *In the Theatre of Consciousness: The Workspace of the Mind*. Oxford: Oxford University Press.

Baars, Bernard J. 1988. *A Cognitive Theory of Consciousness*. Cambridge: Cambridge University Press.

Backhouse, Roger E. 1997. *Truth and Progress in Economic Knowledge*. Cheltenham: Edward Elgar Publishing.
Barkow, J., L. Cosmides and J. Tooby. (eds.) 1992. *The Adapted Mind: Evolutionary Psychology and the Generation of Culture*. New York: Oxford University Press.
Bartlett, Randall. 1989. *Economics and Power: An Inquiry into Human Relations and Markets*. Cambridge: Cambridge University Press.
Bates, Robert H. 1997. "Area Studies and the Discipline: A Useful Controversy." Pp. 166-169 in *PS: Politics and Political Science*. V XXX. N.2.
Batson, C. Daniel. 1991. *The Altruism Question: Toward a Social-Psychological Answer*. Hillsdale, New Jersey: Lawrence Erlbaum Associates.
Becker, Gary. 1987. "Economic Analysis and Human Behavior." Pp. 3-17 in *Advances in Behavioral Economics* V. 1. Ed. by L. Green and J. H. Kagel. Norwood N.J. Ablex Publishing.
Beitz, Charles. 1979. "Bounded Morality: Justice and the State in World Politics." Pp.405-24 in *International Organization*. V. 33.
Bendor, Jonathan and Piotr Swistak. 1997. "The Evolutionary Stability of Cooperation." Pp. 290- 303 in *American Political Science Review*. V.91. N.2
Benecke, R., J. C. Rothwell, J. P. Dick, B. L. Day, and C. D. Marsden. 1987. "Disturbance of Sequential Movements in Patients with Parkinson's Desease." Pp. 361-80 in *Brain*. V. 110.
Berridge, K. C. and J. C. Fentress. 1988. "Disruption of Natural Grooming Chains after Striatopallidal Lesions." Pp. 336-42 in *Psychobiology*. V. 15.
Berridge, K. C. and I. O. Whishaw. 1992. "Cortex, Striatum, and Cerebellum: Control of Serial Order in a Grooming Sequence."Pp. 275-90 in *Experimental Brain Research*. V 90.
Blau, Peter M. 1964. *Exchange and Power in Social Life*. New York: John Wiley.
Blaug, Mark. 1980. *The Methods of Economics: or how economists explain*. Cambridge: Cambridge University Press.
Blessing, William W. 1997. *The Lower Brainstem and Bodily Homeostasis*. Oxford: Oxford University Press.
Blood, Robert. O, Jr. and Donald M. Wolfe. 1960. Husbands and Wives: *The Dynamics of Married Living*. New York: The Free Press.
Boulding, Kenneth. 1950. *A Reconstruction of Economics*. New York: John Wiley.
Bowlby, John. 1988. *A Secure Base. Parent-Child Attachment and Healthy Human Development*. New York: Basic Books.
Dowlby, John. 1969. *Attachment*. Vol.1. New York: Basic Books.
Bromley, Daniel W. 1989. *Economic Interests and Institutions: The Conceptual Foundations of Public Policy*. New York: Basil Blackwell.
Brown, Jason. 1977. *Mind, Brain, and Consciousness*. New York: Academic Press.
Buchanan, James M. 1991. *The Economics and the Ethics of Constitutional Order*. Ann Arbor: University of Michigan Press.
Buchanan, James M and Gordon Tullock. 1962. *The Calculus of Consent: Logical Foundations of Constitutional Democracy*. Ann Arbor: University of Michigan Press.
Burghardt, Gordon M. 1988. "Precocity, Play and the Ectotherm -- Endotherm Transition." Pp.107-48 in *Handbook of Behavioral Neurobiology*, Vol. 9. Ed. by Elliott M. Bass. New York: Plenum.
Burghardt, Gordon M. 1984. "On the Origins of Play." Pp. 5-41 in *Play in Animals and Humans*. Ed. by Peter K. Smith. New York: Basil Blackwell.
Butler, Ann B. and William Hodos. 1996. *Comparative Vertebrate Neuroanatomy: Evolution and Adaptation*. New York: Wiley-Liss.
Campbell, C. B. G. 1992. "Book Review (MacLean: The Triune Brain in Evolution)". Pp. 497-98 in *American Scientist*, V. 80(Sept-Oct 19 1992)
Chertow, Marian R. and D. C. Esty. 1997. *Thinking Ecologically: The next generation of environmental policy*. New Haven: Yale University Press.

Churchland, Patricia S. and Terrence J. Sejnowski. 1992. *The Computational Brain.* Cambridge, Mass: MIT Press.

Clutton-Brock, T. H. 1991. *The Evolution of Parental Care.* Princeton, NJ: Princeton University Press.

Coase, Ronald H. 1960. "The Problem of Social Cost." *Journal of Law and Economics*, 3: 1-44.

Coase, Ronald H. 1937. "The Nature of the Firm." *Economica* 4 (November): 386-405.

Coleman, James S. 1990. *Foundations of Social Theory.* Cambridge, MA: Belknap Press of Harvard University Press.

Cook, Karen S., Jodi O'Brien, and Peter Kollock. 1990. "Exchange Theory: A Blueprint for Structure and Progress," Pp. 158-81 in *Frontiers of Social Theory: The New Synthesis*, Ed. by George Ritzer. New York: Columbia University Press.

Cook, Karen., ed. 1987. *Social Exchange Theory.* Newbury Park, CA: Sage.

Coricelli, Fabrizio and Giovanni Dosi. 1988. "Coordination and Order in Economic Change and the Interpretative Power of Economic Theory." Pp. 124--47 in *Technical Change and Economic Theory.* Ed. by G. Dosi, C. Freeman, R. Nelson, G. Silverberg and L. Soete. London: Pinter.

Corning, Peter A. 1996. "The Cooperative Gene: On the Role of Synergy in Evolution." Pp. 183-207 in *Evolutionary Theory.* V.11.

Corning, Peter A. 1995. "Synergy and Self-organization in the Evolution of Complex Systems. Pp. 89-121 in *Systems Research.* V.12 N.2.

Corning, Peter A. 1983. *The Synergism Hypothesis.* New York: McGraw-Hill.

Cory, Gerald A., Jr. 1998. "MacLean's Triune Brain Concept: in Praise and Appraisal." Pp. 6-19, 22-24. in *Across-Species Comparisons and Psychopathology Society(ASCAP) Newsletter.* V. 11. No. 07.

Cory, Gerald A., Jr. 1997. "The Conflict Systems Behavioral Model and Politics: A Synthesis of Maslow's Hierarchy and MacLean's Triune Brain Concept with Implications for New Political Institutions for a New Century." *Annals of the American Political Science* Association 1997(forthcoming).

Cory, Gerald A., Jr. 1996. *Algorithms, Illusions, and Reality.* Vancouver, WA: Center for Behavioral Ecology.

Cory, Gerald. A., Jr. 1992. *Rescuing Capitalist Free Enterprise for the Twenty First Century.* Vancouver, WA: Center for Behavioral Ecology.

Cory, Gerald A., Jr. 1974. *The Biopsychological Basis of Political Socialization and Political Culture.* Ph.D. Dissertation. Stanford University.

Cosmides, Leda and John Tooby. 1994. "Better than Rational: Evolutionary Psychology and the Invisible Hand." Pp. 327-32 in *American Economic Review.* V. 84. N.2(May).

Cosmides, Leda and John Tooby. 1989. "Evolutionary Psychology and the Generation of Culture, Part II." Pp. 51-97 in *Ethology and Sociobiology.* V. 10.

Crawford, Stephen S. and Eugene K. Balon. 1996. "Cause and Effect of Parental Care in Fishes: An Epigenetic Perspective." Pp. 53-107 in *Parental Care: Evolution, Mechanisms, and Adaptive Intelligence*, ed. by J. S. Rosenblatt and C. T. Snowden. New York: Academic Press.

Crick, Francis. 1994. *The Astonishing Hypothesis: The Scientific Search for the Soul.* New York: Charles Scibner's Sons.

Cromwell, H. C. and K. C. Berridge. 1990. "Anterior Lesions of the Corpus Striatum Produce a Disruption of Stereotyped Grooming Sequences in the Rat." P. 233 in *Society for Neuroscience* Abstracts. V. 16.

Crump, Martha. L. 1996. "Parental Care among the Amphibia." Pp. 109-144 in *Parental Care: Evolution, Mechanisms, and Adaptive Intelligence.* Ed. by J.Rosenblatt and C. T. Snowden,. New York: Academic Press.

Cummings, Jeffrey L. 1993. "Frontal-Subcortical Circuits and Human Behavior." Pp. 873-880 in *Arch Neurol.*. V. 50(Aug).

Cummins, Denise D. 1998. "Social Norms and Other Minds: The Evolutionary Roots of Higher Cognition." Pp. 30-50 in *The Evolution of Mind*. Ed. by D. Cummins and C. Allen. Oxford: Oxford University Press.

Dahlman, Carl, J. 1980. *The Open Field System and Beyond*. Cambridge: Cambridge University Press.

Daly, H. E. (ed.), 1980. *Economics, Ecology, and Ethics*. San Francisco: W. H. Freeman.

Daly, H. E. and Cobb, J. 1989. *For the Common Good: Redirecting the Economy Towards Community, the Environment and a Sustainable Future*. Boston: Beacon Press.

Daum, Irene and Hermann Ackermann. 1995. "Cerebellar contributions to cognition." Pp. 201-10 in *Behavioural Brain Research*. 67.

Davies, James C. 1991. "Maslow and Theory of Political Development: Getting to Fundmentals." Pp. 389-420 in *Political Psychology*. V. 12. N.3.

Davies, James C. 1963. *Human Nature in Politics*. New York: Wiley.

Davis, Lance E. and Douglass C. North. 1970. "Institutional Change and American Economic Growth: a first step towards a theory of institutional innovation." Pp. 131-49 in *Journal of Economic History*. 30.

Devinsky, Orrin and Daniel Luciano. 1993. "The Contributions of Cingulate Cortex to Human Behavior." Pp. 527-56 in *Neurobiology of Cingulate Cortex and Limbic Thalamus: A Comprehensive Handbook*. Ed. by B.Vogt and M. Gabriel. Boston: Birhauser.

Dimitrov, Mariana; J.Grafman, P. Kosseff, J. Wachs, D. Alway, J.Higgins, I. Litvan, J.Lou, and Hallett, Mark. 1996. "Preserved cognitive processes in cerebellar degeneration." Pp. 131-35 in *Behavioural Brain Research*.79.

Dingwall, J. 1980. "Human Communicative Behavior: A Biological Model." Pp. 1-86 in *The Signifying Animal*. Ed. by I. Rauch and G. Carr, G. Bloomington, IN: Indiana University Press.

Donnerstein, Edward and Elaine Hatfield. 1982. "Aggression and Inequity." Pp 309-29 in *Equity and Justice in Social Behavior*. Ed. by Jerald Greenberg and Ronald L.Cohen. New York: Academic Press.

Dugger, William M. 1996. "Four Modes of Inequality." Pp. 21-38 in *Inequality: Radical Institutionalist Views on Race, Gender, Class, and Nation*. Ed. by William M. Dugger. Westport, Connecticut: Greenwood Press.

Durfee, E. H. 1993. "Cooperative Distributed Problem-Solving Between (and within) Intelligent Agents." Pp. 84-98 in *Neuroscience: From Neural Networks to Artificial Intelligence*. Ed. by P. Rudomin, et al. Heidelberg: Springer-Verlag.

Eckel, Catherine C. and Philip Grossman. 1997. "Equity and Fairness in Economic Decisions: Evidence from Bargaining Experiments." Pp.281-301 in *Advances in Economic Psychology*. Ed. by G. Antonides and W. F. van Raaij. New York: John Wiley & Sons.

Edelman, Gerald M. 1992. *Bright Air, Brilliant Fire*. Basic Books.

Eggertson, Thrainn. 1990. *Economic Behavior and Institutions*. Cambridge: Cambridge University Press.

Einstein, Albert. 1954. "Physics and Reality." Pp. 290-323 in *Ideas and Opinions*. New York: Crown Publishing.

Ehrlich, Paul R. and Anne H. Ehrlich. 1996. *Betrayal of Science and Reason*. Wash., D.C. Island Press

Ehrlich, Paul R. and Anne H. Ehrlich. 1990. *The Population Explosion*. New York: Simon and Schuster.

Ehrlich, Paul R. 1969. *The Population Bomb*. New York: Ballentine Books.

Eisenberg, Nancy. 1994. "Empathy." Pp. 247-53 in *Encylopedia of Human Behavior*, edited by V. S. Ramachandran. New York: Academic Press.

Ekins, P., ed., 1986. *The Living Economy: A New Economics in the Making*. London: Routledge & Kegan Paul.

Ekins, P. and M. Max-Neef, (eds). 1992. *Real-Life Economics: Understanding Wealth Creation.* London: Routledge.

Ellis, Ralph. 1986 *An Ontology of Consciousness.* Dordrecht, The Netherlands: Martinus Nijhoff.

Emerson, Richard. 1972a. "Exchange Theory, Part I: A Psychological Basis for Social Exchange." Pp. 38-57 in *Sociological Theories in Progress*, vol. 2. Ed by J. Berger, M. Zelditch, and B. Anderson. Boston, MA: Houghton- Mifflin.

Emerson, Richard. 1972b. "Exchange Theory, Part II Exchange Relations and Networks." Pp.58-87 in *Sociological Theories in Progress*, vol. 2. Ed by J. Berger, M. Zelditch, and B. Anderson. Boston, MA: Houghton- Mifflin.

Eslinger, Paul. J. 1996. "Conceptualizing, Describing, and Measuring Components of Executive Function." Pp. 367-95 in *Attention, Memory, and Executive Function.* Ed.by G. Lyon and N. Krasnegor. Baltimore: Paul H. Brookes Publishing Co.

Etzioni, Amitai. 1988. *The Moral Dimension.* New York: Free Press.

Evensky, Jerry. 1993. "Ethics and the Invisible Hand." Pp. 197-205 in *Journal of Economic Perspectives* 7, No. 2(Spring).

Fagen, R. 1981. *Animal Play Behavior.* New York: Oxford University Press.

Feiwel, George R., editor. 1987. *Arrow and the Ascent of Modern Economic Theory.* London: Macmillan Press.

Feyerabend, Paul K. 1975. *Against Method: outline of an anarchistic theory of knowledge.* London: New Left Books.

Fleischer, S. and B. M. Slotnik. 1978. "Disruption of maternal behavior in rats with lesions of the septal area." Pp. 189-200 in *Physiological Behavior*, 21.

Fleming, Alison S., H. D. Morgan, and C. Walsh. 1996. "Experiential Factors in Postpartum Regulation of Maternal Care." Pp. 295-332 in *Parental Care: Evolution, Mechanisms, and Adaptive Intelligence.* Ed. by J. Rosenblatt and C.Snowden New York: Academic Press.

Fleming, A. S., F.Vaccarino, and C. Leubke. 1980 . "Amygdaloid inhibition of maternal behavior in the nulliparous female rat" Pp. 731-43. in *Physiological Behavior*, 25.

Frank, Robert H., T. Gilovich, and D. Regan. 1993. "Does Studying Economics Inhibit Cooperation? Pp. 159-71 in *Journal of Economic Perspectives* 7. No. 2(Spring).

Frank, Robert H. 1988. *Passions Within Reason: The Strategic Role of the Emotions.* New York: W.W. Norton.

Freeman, John H., Jr.; C. Cuppernell, K. Flannery, and M. Gabriel. 1996. "Limbic thalamic, cingulate cortical and hippocampal neuronal correlates of discriminative approach learning in rabbits." Pp. 123-36 in *Behavioural Brain Research* 80.

Frey. Bruno S. 1992. *Economics as a Science of Human Behavior: Towards a New Social Science Paradigm.* Boston: Kluwer Academic Publishers.

Friedman, Milton. 1953. *Essays in Positive Economics.* Chicago: University of Chicago Press.

Frith, Uta. (1993) 1997. "Autism." Pp. 92-8 in *Scientific American: Mysteries of the Mind.* Special Issue V. 7, N.1.

Firth, Uta. 1989. *Autism: Explaining the Enigma.* Cambridge, MA: Basil Blackwell.

Furubotn, Eirik G. and R. Richter. 1991. "The New Institutional Economics: An Assessment." Pp. 1-32 in *The New Institutional Economics.* Ed. by Furubotn and Richter. College Station, Texas: Texas A&M University Press.

Fuster, Joaquin M. 1997. *The Prefrontal Cortex: Anatomy, Physiology, and Neuro-psychology of the Frontal Lobe.* Third Edition. New York: Lippincott-Raven.

Gagliardo, Anna; F. Bonadonna and I. Divac. 1996. "Behavioural effects of ablations of the presumed 'prefrontal cortex' or the corticoid in pigeons." Pp. 155-62 in *Behavioural Brain Research.* 78.

Galin, David. 1996. "The Structure of Subjective Experience: Sharpen the Concepts and Terminology." Pp. 121-40 in *Toward a Science of Conciousness: The First Tuscon*

Discussions and Debates. Ed, by S. Hameroff, A. Kaszniak, and A. Scott. Cambridge, MA: MIT Press.

Gans, Carl. 1996. "An Overview of Parental Care among the Reptilia." Pp. 145-57 in *Parental Care: Evolution, Mechanisms, and Adaptive Intelligence.* Ed. by J. Rosenblatt and C. Snowden. New York: Academic Press.

Gazzaniga, M.S. 1985. *The Social Brain: Discovering the Networks of the Mind.* New York: Basic Books.

Gill, Flora. 1996. "Comment on Ethics and Economic Science." Pp. 138-56 in *Economics and Ethics?* Ed. by Peter Groenewegen. London: Routledge.

Gloor, Pierre. 1997. *The Temporal Lobe and the Limbic System.* Oxford: Oxford University Press.

Goodall, Jane. 1986. *The Chimpanzees of Gombe: Patterns of Behavior.* Cambridge, Mass.: Harvard University Press.

Goudie, Andrew, (ed.) 1997. *The Human Impact Reader.* Blackwell Publishers.

Gouldner, Alvin. 1960. "The Norm of Reciprocity; a Preliminary Statement." *American Sociological Review* 25: 161-178.

Groenewegen, John, ed. 1996. *Transaction Cost Economics and Beyond.* Boston: Kluwer Academic.

Hall, B. K. 1996. "Homology and Embryonic Development." Pp. 1-37 in *Evolutionary Biology.* V. 28. Ed. by Max K. Hecht, Ross J. MacIntyre, and Michael T. Clegg. New York: Plenum Press.

Hall, B. K., (ed.) 1994. *Homology: The Hierarchical Basis of Comparative Biology.* San Diego: Academic Press.

Hameroff, Stuart R; A. Kaszniak, and A. Scott, (eds.) 1996. *Toward a Science of Conciousness: The First Tuscon Discussions and Debates.* Cambridge, MA: MIT Press.

Hameroff, Stuart R and Roger Penrose. 1996. "Orchestrated Reduction of Quantum Coherence in Brain Microtubules: A Model for Consciousness." Pp. 507-39 in S.Hameroff, A. Kaszniak and A. Scott, eds. 1996. *Toward a Science of Conciousness: The First Tuscon Discussions and Debates.* Cambridge, MA: MIT Press.

Hamilton, W.D. 1964. "The Genetical Evolution of Social Behavior, I & II." Pp.1-16, &17-52 in *Journal of Theoretical Biology.* V. 7.

Hampden-Turner, Charles and Alfons Trompenaars. 1993. *The Seven Cultures of Capitalism.* New York: Doubleday.

Harlow, Harry F. 1986. *From Learning to Love: The Selected Papers of H. F. Harlow.* Ed by C. Harlow. New York: Praeger.

Harlow, Harry F and Margaret K. Harlow. 1965. "The Affectional Systems." Pp. 386-334 in *Behavior of Non-Human Primates.* Ed. by Allen M. Schrier, Harry F. Harlow, and Fred Stollnitz. New York: Academic Press.

Harrington, Anne, (ed.) 1992. *So Human a Brain.* Boston: Birkhauser.

Harth, Erich. 1997. "From Brains to Neural Nets." Pp. 1241-55 in *Neural Networks.* V.10. No.7.

Hatfield, Elaine and Jane Traupmann. 1981. "Intimate Relationships: A Perspective from Equity Theory." Pp. 165-78 in *Personal Relationships. 1: Studying Personal Relationships.* Ed by Steve Duck and Robin Gilmour. New York: Academic Press.

Hayek. Friedrich A. 1991. *Economic Freedom.* Oxford: Basil Blackwell.

Hayek. Friedrich A. 1945. "The Use of Knowledge in Society." Pp. 519-30 in *American Economic Review* (September).

Hayek. Friedrich A. 1937. "Economics and Knowledge." Pp. 33-54 in *Economica*(February)

Heilbroner, Robert L. 1997. "The Embarrassment of Economics." Pp. 18-21 in *Microeconomics 98/99.* Guilford, Conn: Dushkin-McGraw-Hill.

Heilbroner, Robert L. and James K. Galbraith. 1990. *The Economic Problem.* Englewood Cliffs, NJ: Prentice Hall.

Hickman, Cleveland P., Jr., L. Roberts, and F. Hickman. 1990. *Biology of Animals.* Fifth edition. Boston: Times Mirror/Mosby College Publishing.

Hickman, Cleveland P., Jr., L. Roberts, and F. Hickman. 1984. *Integrated Principles of Zoology.* Seventh edition. St. Louis: Times Mirror/Mosby College Publishing.

Hirschman, Albert O. (1982). *Shifting Involvements: Private Interest and Public Action.* New Jersey: Princeton University Press.

Hobhouse, L. T. [1906] 1951. *Morals in evolution: A Study in Comparative Ethics.* London: Chapman & Hall.

Hoffman Elizabeth, K. McCabe, and V. Smith. 1998. "Behavioral Foundations of Reciprocity: Experimental Economics and Evolutionary Psychology." Pp. 335-352 in *Economic Inquiry.* V. 36. N. 3.

Hoffman, M. 1981. "Is Altruism Part of Human Nature?" Pp. 121-37 in *Journal of Personality and Social Psychology,* 40.

Hofstader, Douglas. 1995. *Fluid Concepts and Creative Analogies.* New York: Basic Books.

Holthoff-Detto, Vjera; J. Kessler, K. Herholz, H. Bonner, U. Pietrzyk, M. Wurker, M. Ghaemi, K. Wienhard, R. Wagner, and W. Heiss. 1997. "Functional Effects of Striatal Dysfunction in Parkinson Desease." Pp. 145-150 in *Arch Neurol.* V 54 (Feb).

Homans, George C. 1961. *Social behavior: Its Elementary Forms.* New York: Harcourt Brace and World.

Homans, George C. 1950. *The Human Group.* New York: Harcourt Brace Jovanovich.

Humphrey, N. K. 1976. "The Function of the Intellect." Pp. 303-17 in *Growing Points in Ethology.* Ed. by P.P.G. Bateson and R. H. Hinde Cambridge: Cambridge University Press.

Inglehart, Ronald and Jacques-Rene Rubier. 1978. "Economic Uncertainty and European Solidarity: Public Opinion Trends." Pp. 66-97 in *Annals of the American Academy of Political and Social Science.* V. 440.

Ingrao, Bruna and Giorgio Israel. 1990. *The Invisible Hand: Economic Equilibrium in the History of Science.* Trans. by Ian McGilvray. Cambridge, Mass.: The MIT Press.

Isaac, Glynn. 1978. "The Food-sharing Behavior of Protohuman Hominids." Pp. 90-108 in *Scientific American.* V.238.

Jencks, Christopher. 1990 "Varieties of Altruism." Pp. 53-67 in *Beyond Self-Interest.* Edited by Jane J. Mansbridge. Chicago: University of Chicago Press.

Johnson, Chalmers. 1997. "Perception vs. Observation, or The Contributions of Rational Choice Theory and Area Studies to Contemporary Political Science." Pp. 170-74 in *PS: Politics and Political Science.* V XXX. N.2.

Johnson, David B. 1991. *Public Choice: An Introduction to the New Political Economy.* Mountain View, CA: Mayfield Publishing Company.

Johnson, Glenn L. 1986. *Research Methodology for Economists.* New York: Macmillan.

Kalin, Ned H. 1997. "The Neurobiology of Fear." Pp.76-83 in *Scientific American Mysteries of the Mind.* Special Issue V. 7, N.1.

Kandel, Eric R. James H. Schwartz, and Thomas M. Jessell. 1995. *Essentials of Neural Science and Behavior.* Norwalk, Conn: Appleton & Lange.

Keohane, Robert O. 1990. "Empathy and International Regimes." Pp. 227-36 in *Beyond Self-Interest.* Ed. by Jane J. Mansbridge. Chicago: University of Chicago Press.

Keohane, Robert O. 1989. *International Institutions and State Power.* Boulder, Co: Westview Press.

Kimble, D. P., L. Rogers, and C. Hendrickson. 1967. "Hippocampal lesions disrupt maternal, not sexual behavior in the albino rat." Pp. 401-05 in *Journal of Comparative Physiological Psychology.* 63.

Knutson, Jeanne N. 1972. *The Human Basis of the Polity.* New York: Aldine-Atherton.

Kohlberg, Lawrence. 1984. *The Psychology of Moral Development.* Vol. 2. San Francisco: Harper & Row.

Kohler, Heinz. 1992. *Economics.* Lexington, Mass: D. C. Heath and Company.

Kohler, Heinz. 1968. *Scarcity Challenged: An Introduction to Economics*. New York: Holt, Rinehart and Winston, Inc.
Kokman, Emre. 1991. Book Review(MacLean's The Triune Brain: Role in Paleocerebral Functions.) in *J. Neurosurg*. V.75(Dec)
Krasnegor, Norman A. and Robert S. Bridges, eds. 1990. *Mammalian Parenting: Biochemical, Neurobiological, and Behavioral Determinants*. Oxford: Oxford University Press
Lamendella, J. 1977. "The Limbic System in Human Communication." Pp. 157--222 in *Studies in Neurolinguistics*. Vol. 3. Ed. by H. Whitaker, H. and H. A. Whitaker. New York: Academic Press.
Land, Michael F. and Russell D. Fernald. 1992. "The Evolution of Eyes." Pp. 1-29 in *Annu. Rev. Neurosc*. V. 15.
Langlois, Richard N. 1986a. "The New Institutional Economics: an introductory essay." Pp. 1-15 in *Economics as a Process: Essays in the New Institutional Economics*. Ed. by R.N. Langlois. Cambridge: Cambridge University Press.
Langlois, Richard N. 1986b. "Rationality, Institutions, and Explanation." Pp. 225-55 in *Economics as a Process: Essays in the New Institutional Economics*. Ed. by R.N. Langlois. Cambridge: Cambridge University Press.
LeBerge, D. L. 1995. *Attentional Processing: The Brain's Art of Mindfulness*. Cambridge MA: Harvard University Press.
LeDoux, Joseph E. 1997. *The Emotional Brain*. New York: Simon & Schuster.
Leon, Michael, R. Coopersmith, L. Beasley and R. Sullivan, R. 1990. "Thermal Aspects of Parenting." Pp. 400-15 in *Mammalian Parenting: Biochemical, Neurobiological, and Behavioral Determinants*. Ed. by Norman A. Krasnegor and Robert S. Bridges. Oxford: Oxford University Press.
Leven, Samuel J. 1994. "Semiotics, Meaning, and Discursive Neural Networks." Pp. 65-82 in *Neural Networks for Knowledge Representation and Inference*. Ed. by D. Levine and M. Aparicio. Hillsdale, NJ: Lawrence Erlbaum Associates.
Levi-Strauss, C [1959] 1969. *The Elementary Structures of Kinship*. London: Eyre and Spottiwoode.
Levine, Daniel S. 1986. "A Neural Network Theory of Frontal Lobe Function." Pp, 716--27 in *Proceedings of the Eighth Annual Conference of the Cognitive Science Society*. Hillsdale, NJ: Lawrence Erlbaum Associates.
Lieberman, Philip. 1998. *Eve Spoke*. New York: Norton.
Lorenz, Konrad. 1971. *Studies in Animal and Human Behavior*. Vol. 2. Trans. by R. Martin. Cambridge, MA: Harvard University Press.
Lorenz, Konrad. 1970. *Studies in Animal and Human Behavior*. Vol. 1. Trans. by R. Martin. Cambridge, MA: Harvard University Press.
Losco, Joseph. 1986. "Biology, Moral Conduct, and Policy Science." Pp. 117-44 in *Biology andBureaucracy*. Ed. by Elliott White and Joseph Losco. Lanham, MD: University Press of America.
Lusick, Ian S. 1997. "The Disciplines of Political Science & Studying The Culture of Rational Choice as Case in Point." Pp. 175-79 in *PS: Politics and Political Science*. V XXX. N.2.
MacLean, Paul . D. 1993. "Human Nature: Duality or Triality." Pp. 107-12 in *Politics and the Life Sciences*. V. 12. N.2.
MacLean, Paul . D. 1992. "Obtaining Knowledge of the Subjective Brain("Epistemics")." p.57-70 in *So Human a Brain*. Ed. by Anne Harrington. Boston: Birkhauser.
MacLean, Paul D. 1990. *The Triune Brain in Evolution: Role in Paleocerebral Functions*. New York: Plenum Press.
Maddi, Salvatore, R. 1989. *Personality Theories*. 5th Edition. Chicago, Il: The Dorsey Press.
Malinowski, Bronislaw. 1926. *Crime and Custom in Savage Society*. London: Kegan Paul.

Malinowski, Bronislaw. 1922. *Argonauts of the Western Pacific*. London: Routledge & Kegan Paul.
Mansbridge, Jane, (ed.) 1990. *Beyond Self-Interest*. Chicago: University of Chicago Press.
Margolis, Howard. 1982. *Selfishness, Altruism, and Rationality: A Theory of Social Choice*. Cambridge: Cambridge University Press.
Marsden, C.D. 1984. "Which Motor Disorder in Parkinson's Desease Indicates the True Motor Function of the Basal Ganglia." Pp. 225-41 in *Functions of the Basal Ganglia* (Ciba Foundation Symposium 107). London: Pitman.
Marsden, C.D. 1982. "The Mysterious Motor Function of the Basal Ganglia: The Robert Wartenberg Lecture." Pp. 514-39 in *Neurology*. V. 32.
Maslow, Abraham H. 1968. *Toward a Psychology of Being*. Second Edition. New York: Van Nostrand Reinhold.
Maslow, Abraham H. 1970. *Motivation and Personality*. Second Edition. New York: Harper &Row.
Maslow, Abraham H. 1943. "A Theory of Human Motivation." Pp. 370-96 in *Psychological Review*. V.50.
Masters, Roger D . 1989. *The Nature of Politics*. New Haven: Yale University Press.
Mauss, Marcel. 1954. *The Gift*. Trans. by Ian Cunnison. New York: W.W. Norton.
Maxwell, G. and R. E. Ames, R. E. 1981. "Economists Free Ride: Does Anyone Else?" Pp. 295-310 in *Journal of Public Economists* 15.
Maynard Smith, J. 1982. "The Evolution of Social Behavior - a Classification of Models." Pp. 28-44 in *Current Problems in Sociobiology*. Ed by. King's College Sociobiology Group. Cambridge University Press.
Mazur, Laurie Ann, (ed.) 1994. *Beyond the Numbers*. Wash, D.C.: Island Press.
Miller, Edythe S. 1996. "Seen through a Glass Darkly: Competing Views of Equality and Inequality in Economic Thought." Pp. 87-99 in *Inequality: Radical Institutionalist Views on Race, Gender, Class, and Nation*. Ed. by William M. Dugger. Westport, Connecticut: Greenwood Press.
Miller, Steven A. and John P. Harley. 1992. *Zoology*. Duberque, IA: Wm. C. Brown Publishers.
Miller, Trudi C. 1993. "The Duality of Human Nature." Pp. 221-41 in *Poli tics and the Life Sciences*. V. 12. N.2.
Minsky, M. 1979. "The Social Theory". Pp. 423-50 in *Artificial Intelligence: an MIT Perspective*. Vol. 1. Ed. by P. Winston and R. Brown. Cambridge, MA: MIT Press.
Mirsky, Allan F. 1996. "Disorders of Attention: A Neuropsychological Perspective." Pp. 71-95 in *Attention, Memory, and Executive Function*. Ed. by G. Lyon, G. and N. Krasnegor. Baltimore: Paul H. Brookes Publishing Co.
Mishan, E. J. 1988. *Cost-Benefit Analysis*. 4th edition. London: Unwin Hyman.
Myers, Norman and Julian Simon. 1994. *Scarcity of Abundance? A Debate on the Environment*. New York: W. W. Norton.
Neumann, J. von and O. Morgenstern. 1947. *Theory of Games and Economic Behavior*. Second edition. Princeton, N.J.: Princeton University Press.
Newman, James; B. Baars, and Cho Sung-Bae. 1997. "A Neural Global Workspace Model for Conscious Attention." Pp. 1195-1206 in *Neural Networks*. Vol 10. No. 7.
Newton, Natika. 1996. *Foundations of Understanding*. Amsterdam/Philadelphia: John Benjamins Publishing Co.
Noorderhaven, Niels G. 1996. "Opportunism and Trust in Transaction Cost Economics." Pp. 105-128 in *Transaction Cost Economics and Beyond*. Ed. by John Groenewegen. Boston: Kluwer Academic.
North, Douglass C. 1990. *Institutions, Institutional Change and Economic Performance*. Cambridge: Cambridge University Press.
Nozick, Robert. 1994. "Invisible Hand Explanations." Pp. 314-18 in *American Economic Review*. V.84. N.2(May)

Numan, M., M. J. Numan, and J. B. English. 1993. "Excito-toxic amino acid injections into the medial amygdala facilitate maternal behavior in virgin female rats. Pp. 56-81 in *Horm. Behav.* 27.
Numan, Michael. 1994. "Maternal Behavior." Pp. 221-302 in *Physiology of Reproduction.* 2nd edition. Vol. 2. Ed. by E. Knobil and J. Neill. New York: Raven Press.
Numan, Michael. 1990. "Neural Control of Maternal Behavior," Pp. 231-59 in *Mammalian Parenting: Biochemical, Neurobiological, and Behavioral Determinants.* Ed. by Norman A. Krasnegor and Robert S. Bridges. Oxford: Oxford University Press.
Oberg, R.G.E. and I. Divac. 1979. "'Cognitive' functions of the Neostriatum." Pp. 291-313 in *The Neostriatum.* Ed. by I. Divac and R.G.E. Oberg. Oxford: Pergamon Press.
Parsons, Talcott. 1960. "Pattern Variables Revisited," Pp. 467-88 in *American Sociological Review,* 25.
Parsons, Talcott. 1951. *The Social System.* New York: Free Press.
Penrose, Roger. 1994. *Shadows of the Mind.* London: Oxford University Press.
Peterson, S. A. 1983. "The Psychobiology of Hypostatizing." Pp. 423-51 in *Micropolitics.* 2.
Peterson, S. A. 1981. "Sociobiology and Ideas-Become-Real." Pp. 125-143 in *Journal of Social and Biological Structures.* 4.
Pettman, Ralph. 1975. *Human Behavior and World Politics: A Transdisciplinary Introduction.* London: The MacMillan Press.
Piaget, Jean. 1977. *The Development of Thought.* Trans. by A. Rosen. New York: Viking Press.
Piaget, Jean. [1932] 1965. *The Moral Judgement of the Child.* New York: Free Press.
Pinker, Steven. 1997. *How the Mind Works.* New York: Norton.
Plant, Raymond. 1992. "Enterprise in its place: the moral limits of market." Pp. 85-99 in *The Values of the Enterprise Culture: The Moral Debate.* Ed. by Paul Heelas and Paul Morris. London: Routledge.
Pooley, A.C. 1977. "Nest opening response of the Nile crocodile." Pp. 17-26 in *Journal of Zoology* (London), 182.
Power, Thomas M. 1996. *Environmental Protection and Economic Well-Being.* Second Edition. Armonk, New York: M. E. Sharpe
Postel, Sandra. 1994. "Carrying Capacity: The Earth's Bottom Line." Pp. 48-70 in *Beyond the Numbers.* Wash, D.C.: Island Press
Pribram, Karl H. 1994. "Brain and the Structure of Narrative." Pp. 375-415 in *Neural Networks for KnowledgeRepresentation and Inference.* Ed. by D. Levine and M. Aparicio, M. IV. Hillsdale, NJ: Lawrence Erlbaum Associates.
Pribram, Karl H. 1973. "The Primate Frontal Cortex --Executive of the Brain." Pp. 293-314 in *Psychophysiology of the Frontal Lobes.* Ed. by K. Pribram and A. Luria. New York: Academic Press.
Reiner, Anton, 1990. "An Explanation of Behavior"(review of MacLean's The Triune Brain in Evolution). Pp. 303-05 in *Science,* V. 250 (Oct 12, 1990).
Restak, Richard M. 1994. *The Modular Brain.* A Lisa Drew Book. New York: Charles Scribner's Sons.
Robbins, Lionel. 1952. *An Essay on the Nature & Significance of Economic Science.* Second Edition. London: MacMillan and Co.
Rosenblatt, Jay S. and Charles T. Snowden, (eds.) 1996. *Parental Care: Evolution, Mechanisms, and Adaptive Intelligence.* New York: Academic Press.
Roth, V. Louise. 1994. "Within and Between Organisms: Replicators, Lineages, and Homologues." Pp. 301-37 in *Homology: The Hierarchical Basis of Comparative Biology.* Ed. by Brian K. Hall. New York: Academic Press.
Sagan, Carl. 1977. *The Dragons of Eden.* New York: Random House.
Sahlins, Marshall. 1972. *Stone Age Economics.* Chicago, IL: Aldine - Atherton.

Sahlins, Marshall. 1963. "On the Sociology of Primitive Exchange." Pp. 139-236 *in The Relevance of Models for Social Anthropology*. Ed. by Michael Banton. London: Tavistock Publications.

Salter, Frank K. 1995. *Emotions in Command: A Naturalistic Study of Institutional Dominance.* Oxford: Oxford University Press.

Samuelson, Paul A. 1947. *Foundations of Economic Analysis.* Cambridge, Mass.: Harvard University Press.

Searle, John R. 1997. *The Mystery of Consciousness.* New York: The New York Times Review of Books.

Schmid, A. Allan. 1989. *Benefit Cost Analysis: A Political Economy Approach.*Boulder, Co:Westview Press.

Schmid, A. Allan. 1987. *Property, Power, and Public Choice: An Inquire into Law and Economics.* Second Edition. New York: Praeger.

Schotter, Andrew. 1981. *The Economic Theory of Social Institutions.* Cambridge: Cambridge University Press.

Sen, Amartya K. 1997. *On Economic Inequality.* Expanded Edition. Oxford: Clarendon Press.

Sen, Amartya K. 1979. "Rational Fools: A Critique of the Behavioral Foundations of Economic Theory." in *Philosophy and Economic Theory.* Ed. by F. Hahn and M. Hollis. Oxford: Oxford University Press.

Simmel, Georg. 1950. *The Sociology of Georg Simmel.* Trans. by Kurt H. Wolff. Glencoe, Il: The Free Press.

Simon, H. A. 1948. *Administrative Behavior.* New York: The MacMillan Company.

Simon, H. A. 1972. "Theories of Bounded Rationality." Pp. 161-76 in *Decision and Organization.* Ed. by C. McGuire and R. Radner. Amsterdam.

Simon, Julian. 1981. *The Ultimate Resource.* Princeton, N.J.: Princeton University Press.

Slotnik, B. M. 1967. "Disturbances of maternal behavior in the rat following lesions of the cingulate cortex." Pp. 204-36 in *Behavior*, 29.

Smith, Adam. 1977[1740-90]. "The Correspondence of Adam Smith,(1740-90)", E. Mossner, and T. Ross, eds., in Vol. 6 of *The Glasgow Edition of the Works and Correspondence of Adam Smith.* General editing by D. D Raphael and Andrew Skinner. Oxford: Clarendon Press.

Smith, Adam. 1911[1789]. *The Theory of Moral Sentiments.* New edition. London: G. Bell.

Smith, Adam. 1937[1776]. *The Wealth of Nations.* Ed. by Edwin Cannan. New York: Modern Library.

Smith, C.U.M. 1996 *Elements of Molecular Neurobiology.* 2nd Edition New York: John Wiley & Sons.

Smith, M. Brewster. 1991. "Comments of Davies's 'Maslow and Theory of Political Development." Pp. 421-23 in *Political Psychology.* V.12. N.3.

Solnick, Steven L. 1998. *Stealing the State: Control and Collapse in Soviet Institutions. Cambridge*, Mass: Harvard University Press.

Somit, Albert and Steven A. Peterson. 1997. *Darwinism, Dominance, and Democracy.* Westport, Conn: Praeger Publishers.

Spitz, Rene A. 1965. *The First Year of Life.* New York: International Universities Press.

Stamm, J. S. 1955. "The function of the medial cerebral cortex in maternal behavior in rats." Pp. 347-56 in *Journal of Comparative Physiol. Psychol.* 48.

Stapp, Henry P. 1972. "The Copenhagen Interpretation." Pp. 1098-1116 in *American Journal of Physics.* 40(8).

Steiner, Kurt. (forthcoming). *The Tokyo Trials.*

Stern, Chantal E. and Richard E. Passingham. 1996. "The nucleus accumbens in monkeys (Macaca fascicularis): II. Emotion and motivation." Pp. 179-93 in *Behavioral Brain Research.* 75.

Stigler, George J. 1961. "The Economics of Information." *Journal of Political Economy*. 69(June): 213-15

Strickberger, Monroe W. 1996. *Evolution*. Second Edition. Boston: Jones and Bartlett.

Striedter, Georg F. 1997. "The Telencephalon of Tetrapods in Evolution." Pp. 179-213 in *Brain, Behavior, and Evolution*. 49.

Strum, Shirley S. and Bruno Latour. 1991. "Redefining the Social Link From Baboons to Humans." Pp 73-85 in *Primate Politics*. Ed. by Glendon Schubert and Roger D. Masters. Carbondale and Edwardsville: Southern Illinois University Press.

Stubenberg, Leopold. 1996. "The Place of Qualia in the World of Science." Pp. 41-9 in *Toward a Science of Conciousness: The First Tuscon Discussions and Debates*. Ed. by S.Hameroff, A. Kaszniak and A. Scott Cambridge, MA: MIT Press.

Terlecki, L.J. and R. S. Sainsbury. 1978. "Effects of fimbria lesions on maternal behavior of the rat." Pp. 89-97 in *Physiological Behavior*, 21.

Teske, Nathan. 1997. "Beyond Altruism: Identity-Construction as Moral Motive in Political Explanation." Pp. 71-91 in *Political Psychology*. V.18. N.1.

Thomson, K. S. 1988. *Morphogenesis and Evolution*. Oxford: Oxford University Press.

Thurnwald, Richard. 1932. *Economics in Primitive Communities*. London: Oxford University Press.

Tickell, Chrispen. 1993. "The Human Species: A Suicidal Success." Pp. 219-26 in *Geographical Journal*. Vol. 159.

Tooby, John and Leda Cosmides. 1989. "Evolutionary Psychology and the Generation of Culture, Part I." Pp. 29-49 in *Ethology and Sociobiology*. V. 10.

Tooby, John and I. DeVore, I. 1987 "The Reconstruction of Hominid Behavioral Evolution through Strategic Modeling." in *Primate Primate Models for the Origin of Human Behavior*. Ed. by W. G. Kinsey. New York: SUNY Press.

Trivers, R.L. 1981. "*Sociobiology and Politics*." Pp. 1-44 in *Sociobiology and Human Politics*. Ed. by Elliott White. Lexington , Mass.: D.C. Heath & Company.

Trivers, R.L. 1971. "The Evolution of Reciprocal Altruism." Pp. 35-57 in *The Quarterly Review of Biology*. V. 46.

Trumbo, Stephen T. 1996. "Parental Care in Invertebrates," Pp.3-51 in *Parental Care: Evolution, Mechanisms, and Adaptive Intelligence*. Ed. by J. Rosenblatt and C. Snowden, New York: Academic Press.

Tucker, Don M.; Phan Luu, Karl H. Pribram. 1995. "Social and Emotional Self-Regulation." Pp. 213-39 in *Annals of the New York Academy of Sciences*. Vol. 769.

Tullock, Gordon. 1990. "The Costs of Special Privilege." Pp. 195-211 in *Perspectives of Positive Political Economy*. Ed. by J. Alt, and K. Shepsle. Cambridge: Cambridge University Press.

Tullock, Gordon 1992. *Economic Hierarchies, Organization and the Structure of Production*. Boston: Kluwer Academic Publishers

Udehn. Lars. 1996. *The Limits of Public Choice*. London: Routledge.

Van Valen, Leigh, M. 1982. "Homology and Causes." Pp. 305-12 in *Morphology* 173.

Veblen, Thorstein. 1948[1897]. "Why is Economics not an Evolutionary Science?" Pp. 215-240 in *The Portable Veblen*. Ed. by Max Lerner. Viking Press.

Veenman, C. Leo, L. Medina and A. Reiner. 1997. "Avian Homologues of Mammalian Intralaminar, Mediodorsal and Midline Thalamic Nuclei: Immunohistochemical and Hodological Evidence." Pp. 78-98 in *Brain, Behavior, and Evolution*. 49.

Waal, Frans de. 1996. *Good Natured: The Origins of Right and Wrong in Humans and Other Animals*. Cambridge, MA: Harvard University Press.

Walster, Elaine, G. William Walster, and Ellen Bersheid. 1978. *Equity: Theory and Research*. Boston: Allyn and Bacon, Inc.

White, Elliott. 1992. *The End of the Empty Organism: Neurobiology and the Sciences of Human Action*. Westport, CN: Praeger.

Wilber, Charles K. (ed.) 1998. *Economics, Ethics, and Public Policy*. Lanham, Md: Rowman & Littlefield.

Willhoite, Fred H., Jr. 1981. "Rank and Reciprocity: Speculations on Human Emotions and Political Life." Pp. 239-58 in *Sociobiology and Human Politics*. Ed. by Elliott White. Lexington , Mass.: D.C. Heath & Company.

Williamson, Oliver E. 1996. *The Mechanisms of Governance*. Oxford: Oxford University Press.

Williamson, Oliver E. 1991. "The Logic of Economic Organization." Pp. 90-116 *in The Nature of the Firm: Origins, Evolution, and Development*. Ed. by O. Williamson and Sidney G. Winter. Oxford: Oxford University Press, 1991.

Williamson, Oliver E. 1985. *The Economic Institution of Capitalism*. New York: Free Press.

Williamson, Oliver E. 1975. *Markets and Hierarchies: Analysis and Anti-Trust Implications*. New York: The Free Press.

Wilson, Edward O. 1998. *Consilience: The Unity of Knowledge*. New York: Alfred A.Knopf.

Wispe, Lauren. 1991. *The Psychology of Sympathy*. New York: Plenum Press.

Yankelovich, Daniel. 1981. *New Rules: Searching for Self-Fullfilment in a World Turned Upside Down*. New York: Random House.

Young, H. Peyton. 1994. *Equity: In Theory and Practice*. Princeton, N.J.: Princeton University Press.

Zigler, Edward and Irving L. Child. 1973. *Socialization and Personality Development*. Reading, MA: Addison-Wesley.

Zupan, Mark A. 1998. "Conference on Economics and Sociology." Pp. 333-334 in *Economic Inquiry*. V. 36. N.3.

Index

MIX
Papier aus verantwortungsvollen Quellen
Paper from responsible sources
FSC® C105338

If you have any concerns about our products,
you can contact us on
ProductSafety@springernature.com

In case Publisher is established outside the EU,
the EU authorized representative is:
Springer Nature Customer Service Center GmbH
Europaplatz 3, 69115 Heidelberg, Germany

Printed by Libri Plureos GmbH
in Hamburg, Germany